T0181329

Advances in Information Security

Volume 61

Series Editor

Sushil Jajodia, Center for Secure Information Systems, George Mason University,
Fairfax, VA 22030-4444, USA

For further volumes:
http://www.springer.com/series/5576

Robinson E. Pino • Alexander Kott
Michael Shevenell
Editors

Cybersecurity Systems for Human Cognition Augmentation

Editors
Robinson E. Pino
ICF International
Fairfax, VA, USA

Alexander Kott
Network Science Division
U.S. Army Research Laboratory
Adelphi, MD, USA

Michael Shevenell
ICF International
Darlington, MD, USA

ISSN 1568-2633
ISBN 978-3-319-35222-0 ISBN 978-3-319-10374-7 (eBook)
DOI 10.1007/978-3-319-10374-7
Springer Cham Heidelberg New York Dordrecht London

Printed on acid-free paper

Springer is part of Springer Science+Business Media (www.springer.com)

Preface

This book argues that neuromorphic computing, and particularly the recently emerging efficient hardware architectures for neuromorphic computing, are of particular significance to cyber defense in austere environments and to the continued evolution of any technological roadmap. In this overall volume, we bring examples from government, industry, and academic domains to illustrate special challenges of cyber defense where size, weight, and power (both electric and computing) of devices are tightly constrained. Of special importance are cognitive challenges of cyber users and defenders in austere environments, and to this end we review prior work on augmentation of related cognitive processes, such as visualization and algorithms that attempt to provide context and advice to the human. The state of the art in neuromorphic approaches (such as artificial neural networks) to cyber defense, and their successes and limitations, are discussed along with the emergence of new hardware such as memristor-based computing architectures that opens new opportunities for neuromorphic techniques in cyber defense. This effort sought to pursue, tactical, algorithmic, and hardware approaches currently being pursued within multiple disciplines to advance the state of the art in cybersecurity in particular to human cognition augmentation.

Chapter 1 covers the notion of cyber situational awareness, sensemaking, and situation understanding that are used in the literature to denote different components in the repertoire of cognitive activities exercised by analysts in the prosecution of cyber warfare. The chapter discusses the relative role of these components in cyber analysis and the nature of cognitive challenges they present, focusing on situation understanding. The purpose here is threefold: to clarify the notions, to elevate the role of understanding to that of the key determinant of successful performance, and to offer suggestions for the design of decision aids that are likely to facilitate situation understanding. These issues are tackled from a number of different perspectives. Accordingly, the text is divided into several brief sections that develop a framework and set the stage for design suggestions, and consider the future of intelligent support for cyber warfare predicting transition from "machine learning" to "machine understanding."

In Chap. 2, we present a newly developed in-house neural network learning algorithm called Adaptive Locally Influenced Estimation Network (ALIEN). The aim of this new learning algorithm is to reduce mathematical complexity and electronic overhead in contrast to existing neural network learning models for direct application within physical embedded hardware such as the emerging memristor-, more mature FPGA- or GPU-based technologies for network security applications such as network intrusion detection. In this work, we demonstrate two applications. The first application was to perform malicious network traffic classification within network flows utilizing the often-cited intrusion detection data set from the International Knowledge Discovery and Data Mining Tools Competition. The second application was in the classification of network packets containing DNS queries as A or MX requests. During our experiments, we were able to achieve a 98 % accurate classification of malicious network traffic utilizing only six fields of information and to perfectly classify 20,000 DNS A and MX packets when the training set of only two packets was used (containing one A request and one MX request).

In Chap. 3, we highlight the importance of developing automated tools and models to support the work of security analysts for cyber situation awareness. Current processes are mostly manual and ad-hoc, therefore they are extremely time-consuming and error-prone, and force analysts to seek through large amounts of fine-grained monitoring data, rather than focusing on the big picture of the cyber situation. To address this limitation, we show how an integrated set of automated tools can be used to perform a number of highly repetitive and otherwise time-consuming tasks in a highly efficient and effective way. The result of this type of automated analysis is the generation of a set of higher-level attack scenarios that can be used by analysts to assess the current situation as well as to project it in the near future. We believe this is an important step toward future generations of self-aware and self-protecting systems, but more work needs to be done in this direction to achieve this vision.

In Chap. 4, we focus on data mining in particular to application in modern cyber operations. Defending cyberspace is a complex and largely scoped challenge which considers emerging threats to security in space, land, and sea. Cyberspace is defined as a global domain within the information environment consisting of the interdependent network of information technology infrastructures, including the Internet, telecommunications networks, computer systems, and embedded processors and controllers. And, cyberspace operations are defined as the employment of cyber capabilities where the primary purpose is to achieve military objectives or effects in or through cyberspace. The global cyber infrastructure presents many challenges because of the complexity and massive amounts of information transferred across the global network daily. To this end, we seek to understand the role and practical functionality of data mining.

In Chap. 5, we describe how supply chain threats have invalidated the assumption that one may move critical software out of band of an attacker through the use of secure hardware root of trust. Many systems consist of COTS hardware, which, through supply chain exploitations, may contain trojans. It is no longer valid to assume that an adversary has no reasonable avenue of attack even if the software protections are structured properly and augmented with secure hardware. In today's cyber-attack environment, one must assume that a subset of systems are, or will be, eventually compromised. With this new mindset development of next generation systems should focus on architectures that are capable of supporting design separation for high reliability and information assurance. Furthermore, these systems must be capable of continued operation while under attack and maintain protection of critical intellectual property. By leveraging a hybrid fault model with multiple, parallel execution paths and resultant execution trace comparison, in this chapter, we discuss a distributed architecture where algorithms and applications are fractionated across a cloud computation system to achieve desired security constraints assuring trusted execution. Furthermore, the model architecture can be scaled through proactive thread diversity for additional assurance during threat escalation. The solution provides dynamic protection through distributing critical information across federated cloud resources that adopt a metamorphic topology, redundant execution, and the ability to break command and control of malicious agents.

In Chap. 6, we discuss the future of cybersecurity as a warfare between machine learning techniques of attackers and defenders. As attackers will learn to evolve new camouflaging methods for evading better and better defenses, defense techniques will in turn learn new attacker's tricks to defend against. The better technology will win. Here we discuss the theory of machine learning based on dynamic logic that is mathematically provable to learn with the fastest possible speed. We also discuss cognitive functions of dynamic logic and related experimental proofs. This new mathematical theory, in addition to being provably fastest machine learning technique, is also an adequate model for several fundamental mechanisms of the mind.

In Chap. 7, we focus on malware threats on mobile devices. To address this critical issue, we developed an Artificial Neural Network (ANN)-based malware detection system to detect unknown malware. In our system, we consider both permissions requested by applications and system calls associated with the execution of applications to distinguish between benign applications and malware. We used ANN, a representative machine learning technique, to understand the anomaly behavior of malware by learning the characteristic permissions and system calls used by applications. We then used the trained ANN to detect malware. Using real-world malware and benign applications, we conducted experiments on Android devices and evaluated the effectiveness of our developed system.

In Chap. 8, we describe how the sustainable progress of modern society raises many environmental and organizational issues. Most obvious concerns are related to the problems of energy, as there is no adequate substitute for the depleting hydrocarbons. Especial significance bear energy developments for reliability and security of operational networks. Beyond cyber attacks, the apparent vulnerabilities of the physical integrity of the electrical power grid could be obviated by a decentralized generation of energy. Also, a dependable autonomous supply of energy is decisive for vast distributed networks of sensors and actuators. This chapter reveals a new yet not recognized type of energy in the physical world. Such a possibility could be suspected from many paradoxical observations and experiments where involvement of regular sources of energy is not evident. The surmised new energy is extracted from impetuses of information-processing clocking pulses, the so-called "hot-clocking" effect, which drive the mechanism of the Holographic Universe. Most clearly, this mechanism transpires through the otherwise incomprehensible property of Universe's nonlocality. The considered concept can explain the perplexing "excess heat" effect promising to provide clean abundant energy. This effect had been uneasily attributed to a kind of a nuclear process notoriously dubbed "Cold Fusion" that later on had been largely substituted by a milder term Low Energy Nuclear Reactions (LENR). Proper scientific understanding of the "excess heat" effect would remove the major stumbling block on the way of its reducing to practice.

In Chap. 9, we focus on how the memristor formalism provided by Leon Chua and promoted by Hewlett-Packard Labs has provided a compelling analogy to biological synapses and has led to very rapid progress in the field but it misses much of the complexity that is present in resistive switches. Examining this complexity, it is clear that these devices are in fact much more similar to biological synapses than was previously imagined and a variety of biomimetic opportunities exist for designing neural networks. By leveraging these advanced biomimetic functionalities, the use of memristors in neural networks (and other neuromorphic architectures) shows strong potential as an adaptive and accurate cyberthreat identification solution.

In Chap. 10, we describe how deploying intrusion detection systems (IDS) across all devices in a network can help increase resilience to cyber attacks. Such deployment will require extreme low power hardware to minimize the impact on the power consumption of mobile devices. Several studies have proposed neural network-based IDS. Additionally several other studies have proposed mapping traditional computer algorithms to neural network form to reduce power. This chapter examines the design of several novel specialized multicore neural processors. Such systems could enable pervasive deployment of IDS algorithms. Systems based on SRAM cores and memristor devices were examined. Detailed circuit simulations were used to ensure that the systems could be compared accurately. Two types of memristor cores were examined: digital and analog cores. Novel circuits were designed for both of these memristor systems. Additionally full system evaluation of multicore processors based on these cores and specialized routing circuits were developed. Our results show that the memristor systems yield the highest throughput and lowest power. Our results indicate that the specialized systems can be between two and five orders of magnitude more energy efficient compared to the traditional HPC systems. Additionally the specialized cores take up much less die area—allowing in some cases a reduction from 179 Xeon six-core processor chips to 1 memristor-based multicore chip and a corresponding reduction in power from 17 kW down to 0.07 W.

In Chap. 11, we present a memristor SPICE model and simulation for chalcogenide-based ion-conductor devices. As memristor-based technologies mature, it is important to be able to simulate large numbers of devices within the integrated circuit architecture in order to speed up reliably the development process within the industry standard SPICE simulation environment. Our compact model replicates the characteristic hysteresis behavior through single-valued equations without requiring the need for recursive or numerically intensive solutions. The SPICE model netlist and fitting parameters are presented.

In Chap. 12, we describe the design and operation of a scalable distributed reconfigurable memristor-based computing logic architecture. From a Boolean logic point of view, any computing element functionality can be represented as a truth table that shows completely the validity of the computing logic function. Thus, we have designed and demonstrated the ability to use memristor devices to describe the operation of a distributed functional logic computing architecture. Given that memristor devices are reconfigurable devices whose impedance states are bounded by a maximum and minimum resistance values. Then, with the use of a digital decoder, we can select the distributed memristor device elements which contain the output value of the logic function whose inputs are the digital inputs to the decoder. With this computing architecture scheme any multiple input/output Boolean logic function can be designed and implemented.

In Chap. 13, we talk about how the class of reconfigurable systems, which include the digital field programmable gate array (FPGA) and emerging new technologies such as neuromorphic computation and memristive devices, represents a type of frontier for cyber security. In this chapter, we provide a brief sketch of the field of reconfigurable systems, introduce a few basic ideas about cyber security, and consider the implications of cyber security as it applies to present and future devices. We also attempt to provide some insights on how to add robustness to reconfigurable systems technologies.

Fairfax, VA, USA — Robinson E. Pino
Adelphi, MD, USA — Alexander Kott
Darlington, MD, USA — Michael Shevenell

Contents

Contributors

James B. Aimone Cybersecurity Systems for Human Cognition Augmentation, Sandia National Laboratories, Albuquerque, New Mexico, USA

Massimiliano Albanese Center for Secure Information Systems, George Mason University, Fairfax, VA, USA

Gustave W. Anderson MacAulay-Brown, Inc. (MacB), Roanoke, VA, USA

Simon Berkovich Department of Computer Science, The George Washington University, Washington, DC, USA

Misty Blowers Air Force Research Laboratory, Information Directorate, Rome, NY, USA

Hasan Cam Network Science Division, U.S. Army Research Laboratory, Adelphi, MD, USA

Kristy A. Campbell Department of Electrical and Computer Engineering, Boise State University, Boise, ID, USA

George Corbin BAE Systems, Rome, NY, USA

Erik P. Debenedictis Cybersecurity Systems for Human Cognition Augmentation, Sandia National Laboratories, Albuquerque, New Mexico, USA

Arthur Edwards U.S. Air Force Research Laboratory, Kirtland, NM, USA

Stefan Fernandez Air Force Research Laboratory, Information Directorate, Rome, NY, USA

Brandon Froberg Air Force Research Laboratory, Information Directorate, Rome, NY, USA

Xinwen Fu Department of Computer Science, University of Massachusetts, Lowell, MA, USA

Linqiang Ge Department of Computer & Information Sciences, Towson University, Towson, MD, USA

Raqibul Hasan University of Dayton, Dayton, OH, USA

Sushil Jajodia Center for Secure Information Systems, George Mason University, Fairfax, VA, USA

Alexander Kott Network Science Division, U.S. Army Research Laboratory, Adelphi, MD, USA

Andrew J. Lohn Cybersecurity Systems for Human Cognition Augmentation, Sandia National Laboratories, Albuquerque, New Mexico, USA

James Lyke U.S. Air Force Research Laboratory, Kirtland, NM, USA

Matthew J. Marinella Cybersecurity Systems for Human Cognition Augmentation, Sandia National Laboratories, Albuquerque, New Mexico, USA

Mark R. McLean Center for Exceptional Computing, Baltimore, MD, USA

Patrick R. Mickel Cybersecurity Systems for Human Cognition Augmentation, Sandia National Laboratories, Albuquerque, New Mexico, USA

Kevin Nelson BAE Systems, Rome, NY, USA

Antonio S. Oblea Department of Electrical and Computer Engineering, Boise State University, Boise, ID, USA

Leonid Perlovsky LP Information Technology & Harvard University, Cambridge, MA, USA

Youngok K. Pino Information Sciences Institute, University of Southern California, Arlington, VA, USA

Robinson E. Pino U.S. Department of Energy, Office of Science, Washington, DC, USA

Olexander Shevchenko LP Information Technology, Cambridge, MA, USA

Tarek M. Taha University of Dayton, Dayton, OH, USA

Jonathan Williams Air Force Research Laboratory, Information Directorate, Rome, NY, USA

Guobin Xu Department of Computer & Information Sciences, Towson University, Towson, MD, USA

Chris Yakopcic University of Dayton, Dayton, OH, USA

Wei Yu Department of Computer & Information Sciences, Towson University, Towson, MD, USA

Yan Yufik Institute of Medical Cybernetics, Inc., Potomac, MD, USA

About the Editors

Robinson E. Pino is a visionary thinker and technology leader in cybersecurity, computational intelligence, computing architectures and development with the current role of Program Manager for the Advanced Scientific Computing Research (ASCR) program office in the Department of Energy's (DOE) Office of Science. In his portfolio, Dr. Pino focuses on revolutionary basic research and development efforts for high performance computing, cybersecurity, and applications that will enable our continued leadership through exascale and beyond computing and energy efficient technologies. Dr. Pino has deep and broad expertise within technology development, program management, government, industry, and academia. He previously worked as Director of Cyber Research at ICF International advancing the state of the art in cybersecurity by applying autonomous concepts from computational intelligence and neuromorphic computing for the U.S. Department of Defense (DoD) Army Research Laboratory (ARL) and various DoD and U.S. Department of Energy (DoE) collaborators, industry and academia. Dr. Pino's research and development program focused on the development of intelligent, autonomous, and cognitive applications toward network, host, and mobile security solutions. In addition, Dr. Pino was a Senior Electronics Engineer at the U.S. Air Force Research Laboratory (AFRL) where he was a program manager and principle scientist for the computational intelligence and neuromorphic computing research efforts. He also worked at IBM as an advisory scientist/engineer development enabling advanced CMOS technologies and as a business analyst within IBM's photomask business unit. Dr. Pino served as an adjunct professor at the University of Vermont where he taught electrical engineering courses. Dr. Pino has a Ph.D. and M.Sc. degrees in Electrical Engineering with honors from the Rensselaer Polytechnic Institute and a B.E. in Electrical Engineering with honors from the City University of New York, City College. He is the recipient of numerous awards and professional distinctions; has published more than 50 technical papers, including four books; and holds six patents, three pending.

Alexander Kott is responsible for fundamental research and applied development in performance and security of tactical mobile and strategic networks. He oversees projects in network performance and security, intrusion detection, and network emulation. Research under his direction brings together government, industry, and academic institutions working toward a fundamental understanding of interactions, interdependencies, and common underlying science among social/cognitive, information, and communications networks, including science for cyber. Previously, Dr. Kott served as a program manager for the Defense Advanced Research Programs Agency (DARPA), and earlier positions included technical director with BBN Technologies, director of R&D at Logica Carnegie Group, and IT research department manager at AlliedSignal, Inc. In 2008, he received the Secretary of Defense Exceptional Public Service Award and accompanying Exceptional Public Service Medal. Dr. Kott has a Ph.D. from the University of Pittsburgh. He has published more than 70 technical papers and coauthored and edited six technical books.

Michael Shevenell has over 25 years of experience managing and implementing the development of network applications using several technologies, including Python, Java, C++, Perl, PHP, CORBA, and C. He is also experienced in database development (Greenplum, PostgreSQL, Oracle, and MYSQL). Mr. Shevenell has led and managed teams of software developers and researchers that design and implement cybersecurity research, development, and applications. He has experience in the development of applications that take advantage of various types of Linux file-systems, cluster environments, and network emulation systems. Mr. Shevenell has played a major role in the research, design, and development of advanced Intrusion Detection System (IDS).

Chapter 1
Situational Awareness, Sensemaking, and Situation Understanding in Cyber Warfare

Yan Yufik

1.1 Introduction

The notions of situational awareness, sensemaking, and situation understanding are used in the literature to denote different components in the repertoire of cognitive activities exercised by analysts in the prosecution of cyber warfare. This chapter discusses the relative role of these components in cyber analysis and the nature of cognitive challenges they present, focusing on situation understanding. The purpose is threefold: to clarify the notions, to elevate the role of understanding to that of the key determinant of successful performance, and to offer suggestions for the design of decision aids that are likely to facilitate situation understanding. These issues are tackled from a number of different perspectives. Accordingly, the text is divided into six brief sections: Sects. 1.1–1.4 develop a framework and set the stage for design suggestions in Sect. 1.5. Section 1.6 considers the future of intelligent support is cyber warfare predicting transition from "machine learning" to "machine understanding." (Throughout the chapter, the terms "situation comprehension" and "situation understanding" will be used interchangeably).

1.2 Discussing the Terms

Review [2] postulates that situational awareness (SA) for cyber defense encompasses the following seven aspects:

"1. Be aware of the current situation.
2. Be aware of the impact of the attack.

Y. Yufik (✉)
Institute of Medical Cybernetics, Inc., Potomac, MD 20854, USA
e-mail: imc.yufik@att.net

R.E. Pino et al. (eds.), *Cybersecurity Systems for Human Cognition Augmentation*, Advances in Information Security 61,
DOI 10.1007/978-3-319-10374-7_1

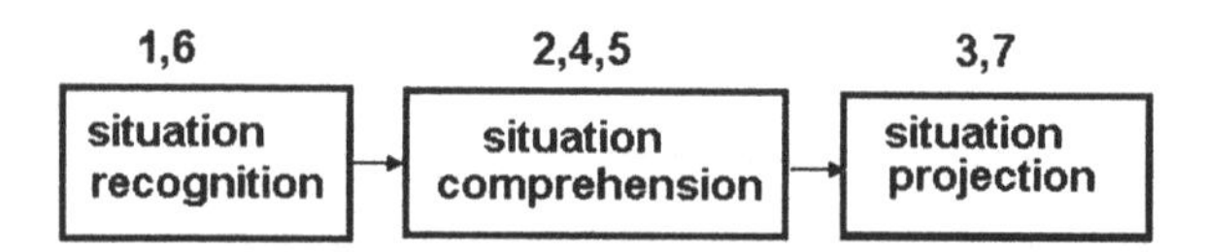

Fig. 1.1 The SA situation recognition (including Aspects 1 and 6), situation comprehension (including aspects 2, 4, and 5), and situation projection (including aspects 3 and 7) (adopted from [2])

3. Be aware of how situations evolve.
4. Be aware of actor (adversary) behavior.
5. Be aware of why and how the current situation is caused.
6. Be aware of the quality (and trustworthiness) of the collected situation awareness information items and the knowledge-intelligence-decisions derived from these information items.
7. Assess plausible futures of the current situation" (Barford et al. [2], pp. 3–4).

These aspects combine into three stages of SA, as shown in Fig. 1.1.

According to Fig. 1.1, comprehending a situation involves awareness of the perpetrator, the cause, the means and the impact. An extensive quote below presents a less general and more technical definition of comprehension tailored to the cyber domain:

> *SA Stage 2*: *Comprehension*. During the second cognitive fusion stage (situation assessment), an analyst performing escalation analysis first reviews the suspicious activity (perception). The analyst then combines and integrates his knowledge and experience with additional data sources to determine whether the suspicious activity represents an actual incident. He refines the mental model of the attacker's identity and threat level as he traces the attacker's path through the network back in time. Also in this stage, an analyst performing correlation analysis identifies and reports on patterns of suspicious and anomalous behavior; the reports serve to cue other analysts. This escalation and correlation analysis represents the comprehension aspect of SA. SA Stage 2 also involves limited projection. Escalation analysis includes some postulating about an attacker's actions if left unblocked. Incident responders, in choosing a course of action, project what future actions an unblocked attacker could take and what actions an attacker might take if he realizes he has been discovered ([4], pp. 229–233).

Per that definition, comprehending a situation in the cyber battlespace boils down to escalation and correlation analysis. Figure 1.2 places situation comprehension within a decision loop.

The process in Fig. 1.2 is, to an extent, consistent with the following definition of sensemaking:

> Sensemaking, as in to make sense, suggests an active processing of information to achieve understanding (as opposed to the achievement of some state of the world), and this is sense in which we mean it here: Sensemaking involves not only finding information but also requires learning about new domains, solving ill-structured problems, acquiring situation awareness, and participating in social exchanges of knowledge. In particular, the term encompasses the entire gamut of behavior surrounding collecting and organizing information for deeper understanding ([22]).

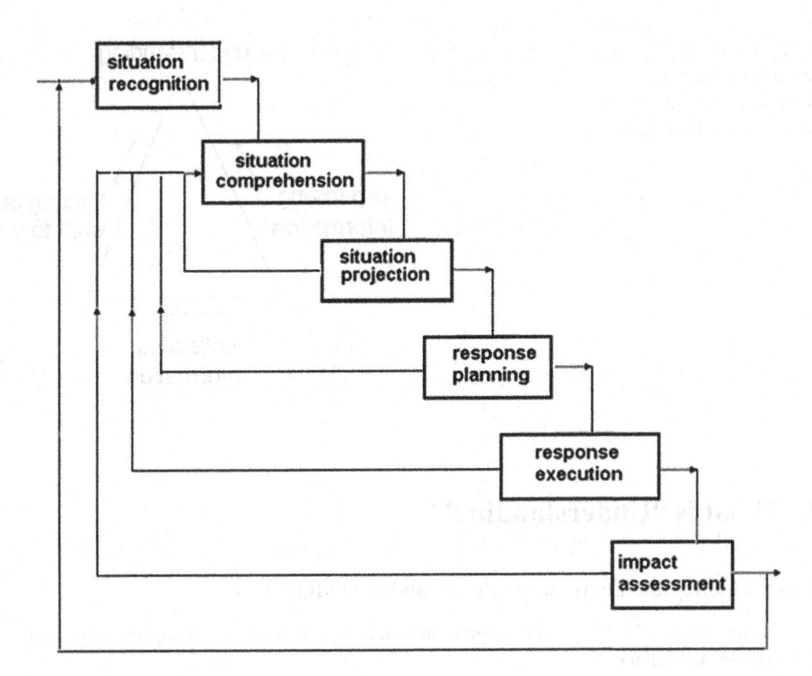

Fig. 1.2 Envisioning how the situation can evolve and formulating appropriate responses are predicated on reaching an adequate level of situation comprehension. The diagram emphasizes that the process is not linear: Comprehension can improve in the course of cognitive activities in every stage, which, in turn, can lead to adjustments in those activities. Arrival of new data might or might not change the way the analyst views the current situation

According to the definition, "sensemaking" denotes cognitive activities directed towards and preparing the stage for the onset and subsequent improvement of understanding. (Note discrepancies in the interpretations of the concepts: Situational awareness can be viewed as a constituent of sensemaking [22, 16] or the other way around [2]. This chapter adheres to the former interpretation). Collecting and organizing information culminates in achieving understanding, as illustrated in Fig. 1.3.

Figure 1.3 connotes several ideas:

1. Information processing necessary for reaching understating, is a laborious activity.
2. The fruits of the labor include fast and streamlined action planning.
3. The apex of understanding might never be reached (or reached too late, in the fashion of Monday-morning quarterbacking).

Reaching the apex (grasp) is accompanied by the emergence of cohesive cognitive structures (mental models) amenable to cognitive operations not available on the sets of disjointed information elements. Efficient decision making is the product of such operations (for more on this subject, please see [27–32]).

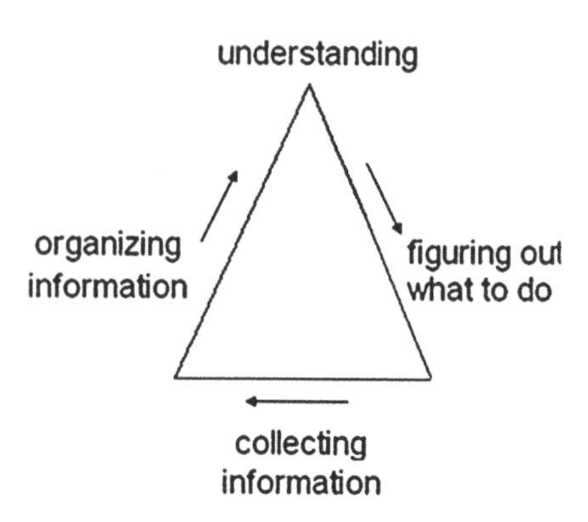

Fig. 1.3 Situation understanding mediates between information gathering and action planning

1.3 What is "Understanding"?

Webster's Collegiate Dictionary defines understanding as

> 1: a mental grasp …2a: the power of comprehension, *esp*: the capacity to apprehend general relations of particulars.

Apprehending general relations of particulars is the hallmark of human cognition (e.g., the Latin *cogito* derives from *coagitare*, to shake together ([13], p. 183)). "Shaking together" some disjointed information elements can result in either piling them into a heap or creating a well-organized, cohesive cognitive structure (mental model). Figure 1.4 illustrates the distinction.

The job of climbing the pyramid (as in Fig. 1.3), from the bottom of information gathering to the summit of understanding, is the sole responsibility of the decision-maker. Assume that the driver takes the trouble of collecting information beyond the minimum needed for reaching the destination from the current location and then organizing this information into a coherent model, what would such an effort afford? The results include, but are not limited, to the following:

1. Ability to figure out and flexibly adjust routes between any points (as opposed to learning some fixed routes and rigidly adhering to them).
2. Ability to retain different routes in one's memory and retrieve them on demand.
3. Ability to reverse one's mental steps (i.e., going back from B to A in one's mind) and trace possible alternative routes, without having those traces interfere with each other.
4. Ability to deal efficiently with unforeseeable obstacles such as a street that is closed off because of an accident.
5. Ability to determine whether novel inputs or, at least, some of them are relevant (salient) or irrelevant to the task at hand (e.g., the sign "street is closed" is relevant if the street happens to be on the current route and irrelevant if otherwise).
6. Ability to foresee the forthcoming inputs (or, at least, some of them, as in "a second roundabout will have to come up shortly after passing the first one").

Fig. 1.4 Driving in the pre-GPS era. The officer understands the situation (holds a well-organized mental model) while the driver does not. Instructions can make the driver aware of the relevant information elements (the "particulars") but cannot do the job of pulling those elements together into an actionable model. (Drawing by Stevenson, 1976, The New Yorker Magazine, Inc.)

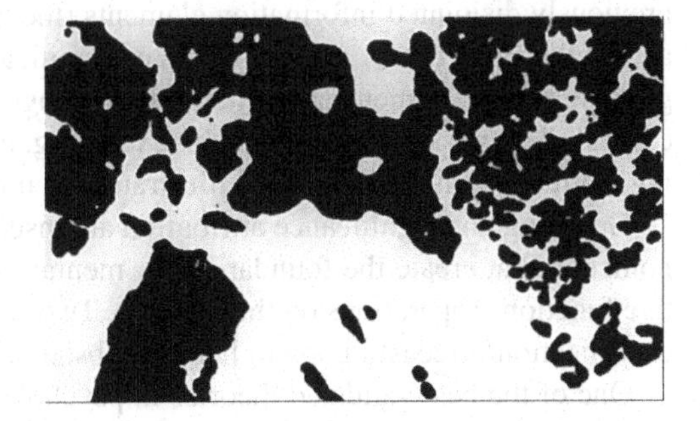

Fig. 1.5 An organized structure (image of a bearded man) might or might not emerge from the aggregation of shapeless blobs. It was found that some people fail to grasp the image, however hard and for however long they try

The latter three abilities become crucial when time is of the essence. Since in real-life situations time is always of the essence, albeit with different degrees of urgency, the key dilemma facing decision-makers consists of choosing between (a) expending time and effort in hope of reaching situation understanding thereby enabling robust performance and (b) avoiding such expenditures and thus increasing the risk of performance breakdown. Two factors exist that can bias a decision-maker toward the latter choice: Expenditures can be steep and success is never certain. An image in Fig. 1.5 ubiquitous in psychology textbooks will help to appreciate these factors.

In difficult tasks, the temptation is to constrain oneself to seeking local relations and abandon looking for the global ones, as in Fig. 1.6.

Fig. 1.6 The effort of organizing information can be extended to a few available information elements while ignoring the rest (e.g., some people are quick to see a fig. of a person throwing an object with the right hand)

1.4 Anatomy of Grasp

Two transformations appear to occur simultaneously at the apex in Fig. 1.4: The previously disjointed information elements (the "particulars") form distinct groups and, at the same time, different levels of significance (weights) are attributed to the groups and the elements. It is helpful to recognize that these transformations are expressed in two strategies of machine learning: unsupervised and supervised learning, correspondingly. Figure 1.7 illustrates the notion.

Grouping and significance attribution are inseparable facets of the grasping phenomenon that create the foundation for mental models and enable complex cognitive functions (operations on the models). Two examples—similarity judgment and classification/forecasting—will help to substantiate these contentions.

One of the best validated theories in psychology determines similarity S (A, B) of two objects A and B as a function F of the number of shared attributes (features) minus the number of distinctive attributes, as follows:

$$S(A,B) = V_1F(a \cap b) - V_2F(a-b) - V_3F(b-a)$$

Here a and b are lists of features of A and B, $(a \cap b)$ is the overlap, (a-b) and (b-a) are the features of A and B identified as distinctive when comparing A to B and B to A, correspondingly (similarity determinations are not symmetrical), and V_1, V_2 and V_3 are parameters (weights) reflecting subjective significance attributed to various features [26]. Besides accounting for subjective significance, the theory postulates prior extraction of relevant features ("…the representation of an object as a collection of features is viewed as a product of a prior process of extraction and compilation" ([26], p. 329). That is, the number of features one can associate with an object

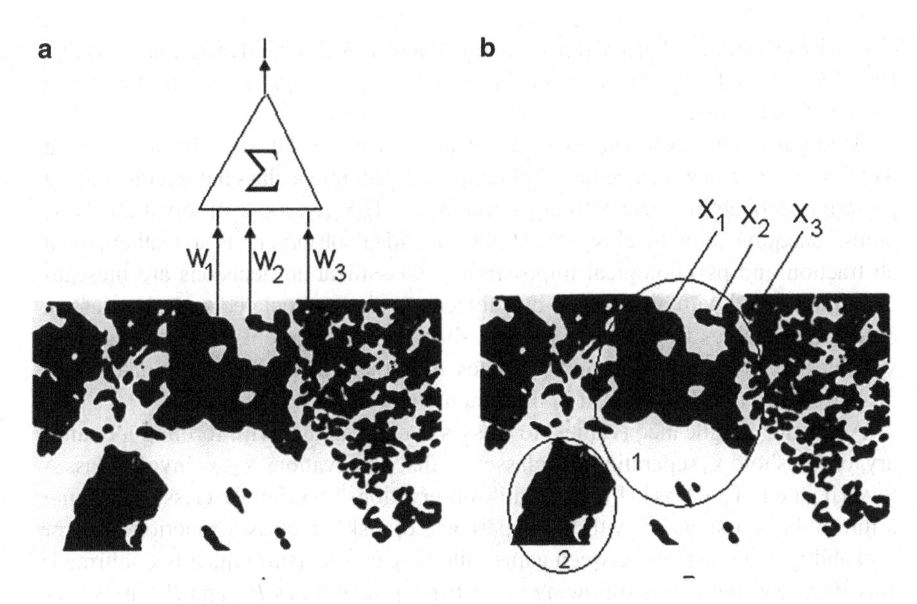

Fig. 1.7 Supervised learning (e.g., in neural nets) attributes relative weights to the input elements (**a**) while unsupervised learning groups those elements based on some clustering criteria (**b**). The image is comprised of group 1 (head of a bearded man looking straight at the viewer) and group 2 (his right upper arm). Elements x_1 and x_2 are members of group 1 (the eyes) while element x_3 is not. Learning algorithms assign different weights to these elements and different strength to interactions between them (elements in a cluster interact more strongly with each other than with the residents of other clusters. Note that x_2 is almost equidistant from x_1 and x_3 but x_1 is inside cluster 1 while x_3 is outside. This suggests that spatial proximity is not the clustering criterion or, at least, not the only one)

is practically unlimited which makes similarity determination impossible unless a manageable subset of features is extracted from the indefinitely large set. Extraction is concomitant with attributing high significance to some features and low significance to the other ones. For example, one can assert that "Washington is like Paris" on the grounds that both cities have radial street layout ($a \cap b$) although Washington does not have as many French restaurants (a-b) and Paris has a vibrant nightlife while Washington does not (b-a). This informed judgment has selected for comparison three urban characteristics out of an infinite number of other possibilities and has favored radial design above all the other attractions.

The dual process of feature selection and value attribution both enables classification and forecasting and strongly influences their outcomes. Clearly, classifying objects is contingent on first selecting some finite subsets of features out of indefinitely large sets, to be subsequently used in the classification decisions. The same holds for forecasting: It cannot proceed until a sufficiently narrow subset of features is selected to be treated as predictors. Hence, the enabling.

Let classification of objects A and B rely on a single feature, or variable x which assumes values in some interval X_A in the case of A and in the interval X_B in the case of B. The probability density functions for A and B over x are denoted as $f_A(x)$ and $f_B(x)$, correspondingly. Overlapping intervals, $X_A \cap X_B \neq \phi$, as in Fig. 1.8, are a source of ambiguity.

Ambiguity is not experienced by an "ideal" decision-maker (i.e., the one meeting Kotelnikov criteria which demands placing an object into a class having the highest posteriori probability. Under the criteria, since $f_B(x_0) > f_A(x_0)$, observation $x = x_0$ points unequivocally to class B). However, "ideal observer" is a mathematical abstraction and psychological impossibility: Classification decisions are inescapably intertwined with and shaped by subjective values associated with the classes. Selecting variables is, in and of itself, a sharp value judgment that elevates some variables to the foreground and relegates the others to the indiscriminate background (an issue I shall re-visit in the concluding remarks).

Ascribing significance (values) to classes is concomitant with forming a boundary, or threshold x_0 separating the classes so that observations $x \leq x_0$ invoke class A (even if $f_B(x_0) > f_A(x_0)$, as in Fig. 1.4) while observations $x > x_0$ invoke class B. Forming a threshold is associated with the experience of risk which is a function of error probability and trade-off between gains following correct (subsequently confirmed) classifications and losses following error. Error probabilities P_A^* and P_B^* are sensitive to the position of the boundary x_0

$$P_A^* = \int_{x_0}^{\infty} f_A(x)\,dx; \quad P_B^* = \int_{-\infty}^{x_0} f_B(x)\,dx$$

Minimal risk over a sufficiently large number of observations and classifications episodes is obtained by positioning boundary x_0 according to the ratio

$$x_0 = \frac{P_A^*\left(h_A^* - h_A\right)}{P_B^*\left(h_B^* - h_b\right)}$$

where h_A^* is the cost of misclassifying A, h_A is the reward following correct classification, h_B^* and h_b are the cost and the reward in the case of B, correspondingly (e.g., Helstrom [11]).

Classification delivers no intrinsic bonuses to the organism except in the service of risk mitigation. It is not unreasonable to assume that biological classification mechanisms have been chiseled by evolution to deliver optimal risks while keeping efforts needed to reduce error under the present circumstance (by increasing the number and varying the content of observations) within limits commensurate with the history of prior experiences. Stated differently, in classification and forecasting decisions, feature probabilities and correlations are taken into account in an interplay with other factors receiving summary expression in subjective values.

Value adjustments cause repositioning of the separation threshold (see Fig. 1.8). As a result, values dominate classification decisions and, in the limit, can overrule the influence of probabilities. Curiously, the Bayes' formula which takes into

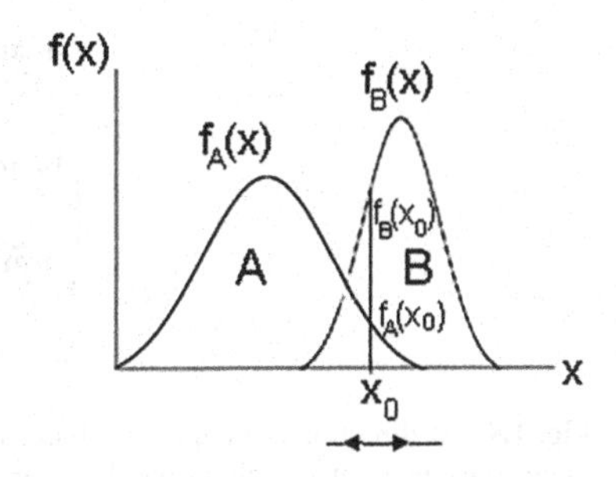

Fig. 1.8 Overlapping probability density functions engender ambiguity (i.e., $x = x_0$ allows either A or B)

account only probabilities reveals also the margins where they no longer matter. An a priori decision devaluing class A shifts the threshold towards $x_0=0$ which deems any observation $x=x_i$ irrelevant and thus gives no additional support to hypothesis H(A), H(A)=0. As a result, no number of observations of any nature $x=x_i$ might succeed in changing the posteriori, $H(A/x_i)=0$. Inversely, overvaluing class A shifts the threshold in the opposite direction which sets H(A) at H(A)=1 and allows any observation $x=x_i$ to be construed as supporting A, $H(x_i/A)/H(x_i)=1$. As a result, again, no number of observations of any nature will have a chance of changing the posteriori, $H(A/x_i)=1$.

To conclude this section, an example from chess theory will be used to stress again the duality of feature grouping and values attribution in attaining grasp. Chess theory attributes different values to pieces, based on the number of chessboard squares they can control. For example, the queen placed in the center of the board (d4) controls 27 squares and is given 9 value units while the rook (the next strongest piece) controls only 14 squares from any position and is given 5 units, etc. These values hold, in a sense, until the game begins: As the pieces start moving, they form groups curtailing the zones of control. As observed by the former world champion:

> The piece values fluctuate depending on your position and can change at every turn… Calculation in chess is not like one plus one, it's more like figuring out a route on a map that keeps changing before your eyes ([12], pp. 48, 87).

Grouping and attributing values to pieces and strength to interactions are inextricably intertwined. Cyber warfare is more complex than chess in many ways. For example, your chess opponent must wait until you make a move while cyber opponents owe you no such courtesies.

1.5 Apprehending Global Relations in a Multitude of Particulars

A situation can involve multiple elements entangled in a multitude of local interactions. Identifying those is the subject of information organization on the way to the apex (see Fig. 1.4). At the apex, two distinct integrative entities are formed along with a relation between them. For example, situations "Romeo loves Juliette",

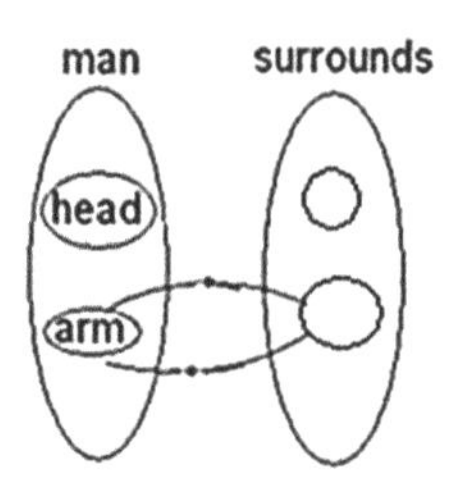

Fig. 1.9 Bi-directional mapping includes coordinated and reversible changes in the composition of entities, the relative values and the strength of interaction between the entities and between elements inside the entities. One form of such mapping involves reversible transfer of elements across entity boundaries (e.g., group 2 in Fig. 1.7 can be seen alternatively as a part of the image or a part of the surrounds). Transfers are accompanied by value and strength adjustments (e.g., value is degraded when switching from "upper arm" in "man" to a blob in the "surrounds" and restored on the reversal; interaction between "arm" and "head" in "man" is stronger than the interaction between blobs in the surrounds)

"USA trades with China," "whites have advantage in the center," "person A is receiving directions from person B" and "a bearded man stands out from indiscriminate background" are all comprised of two entities (the environment can be one of them). Entities are integrative in the sense that they can be comprised of many groupings, with an unlimited degree of nesting (e.g., (surrounds, man (arm, head (eyes, nose, hair (beard, ...)))).

Relation is a coordination mechanism: Changes in one entity both necessitate and constrain changes in the other one. One way to describe the mechanism is to represent it as bidirectional mapping, as in Fig. 1.9.

In the movie "Phenomenon," the protagonist receives a boost to his IQ after an encounter with extraterrestrials. The man's newly bestowed intellectual abilities are applied to a problem that had defied him for a long time: his vegetable garden was invaded regularly by a rabbit bent on destroying the crop. A fence was built to keep the rabbit out and fortified many times, all in vain. Finally, the elevated IQ revealed solution: the rabbit resided in the garden so fortifications intended to keep it out were actually keeping the rabbit in. There are three points here: First, understanding involves re-grouping; second, re-grouping demands effort because groups, once formed, tend to persist and, third, the requisite amount of effort might or might not be available.

Bipartite cognitive structures comprised of integrative entities and a relation mechanism performing reversible and coordinated operations on the entities and their constituent groupings form "mental models" that underlie situation understanding. Understanding enables semantic forecasting (prediction and retrodiction) and simplifies and streamlines action planning (see Fig. 1.3). Semantic forecasting is different from statistical forecasting in that the former emanates from apprehending relations while the latter builds on detecting correlations. For example, the image in Fig. 1.4 can give rise to models "person A receiving directions from person B" or "person is wearing a hat", both models allow semantic forecasting although of different depth and scope (i.e., relation "A wears B" predicts and retrodicts co-location of A and B"). By contrast, a juxtaposition "a man, a hat" asserts no relation and thus does not constitute a model. Accordingly, no semantic forecasting is enabled while statistical forecasting is still possible providing recurrence of the juxtaposition has been observed a sufficient number of times. In a similar way, relations "A is fenced in/out by B" predict and retrodict future and past whereabouts of B respective A. Relation "white rook attacks black king" attributes high value to the rook and high strength to the interaction between the pieces, and it predicts the repertoire of responses available to Black and retrodicts the range of moves that could plausibly precede the last one. Relations can be expressed as functions (which are not necessarily computable).

1.6 Own Assets A are Attacked by Exploits B

Cyber battles are similar to conventional ones in that both aim to protect own assets while incapacitating and/or degrading those of the opponent. The adversaries control limited amounts of offensive and defensive resources (exploits and countermeasures), which they strive to deploy in a manner that maximizes risks to the opponent while minimizing exposure of their own assets. There are also significant differences, the most prominent of which is a higher degree of interdependence between activities in a cyber network as opposed to that in a conventional battle space. "The objects of research in cyber security are:

- Attacker A along with the attacker's tools (especially malware) and techniques T_a.
- Defender D along with the defender's defensive tools and techniques T_d and operational assets, networks and systems N_D.
- Policy P, a set of defender's assertions or requirements about what events should and should not happen. To simplify, we may focus on cyber incidents I: events that should not happen. .... A shorthand for the totality of such relation may be stated as

$$(I, T_d, N_d, T_a) = 0 \qquad ([14], p.2)$$

Consistent with the above definition, global relations in the cyber battle space can be defined as follows [29]:

The network contains asset composition A comprised of S own assets. Assets A are attacked by N exploits forming attack composition B. Different exploits can cause different degrees of degradation in different assets. An analyst controls a repertoire of M countermeasures that can remove exploits or reduce the degradation they cause. Tools for detecting and analyzing exploits are less than perfect so an analyst cannot always be certain about either the impact the exploits will have on the assets or the impact the countermeasures will have on the exploits. The objective is to minimize the overall expected degradation inflicted by attack composition B in the asset composition A. The problem can be expressed as follows:

$$F(\|\lambda\|)=\sum_{i=1}^{S}V_i(1-\prod_{j=1}^{S}(1-\{1-\prod_{\mu=1}^{N}\{1-\varpi_{\mu j}\times \\ \times\prod_{h=1}^{N}(1-[1-\prod_{r=1}^{M}q_{rh}^{\lambda_{rh}}]\beta_{h\mu})\}\}\alpha_{ji}))\rightarrow \min \tag{1.1}$$

under constraints

$$V_i\geq 0;\ \ \sum_{h=1}^{N}\lambda_{rh}=1,\ \ r=1,2,\ldots M,\ \ \lambda_{rh}\in\{0,1\},\ \ 0\leq(q_{rh}=1-p_{rh})\leq 1$$

$$0\leq(\varpi_{\mu j}=1-\varepsilon_{\mu j})\leq 1;\ \ 0\leq\alpha_{ji}\leq 1;\ \ 0\leq\beta_{h\mu}\leq 1;\ \ h,\mu=1,\ldots,N;\ \ i,j=1,\ldots,S$$

Here V_i is the relative significance (value) of the ith asset, $\varpi_{\mu j}$ is the degree of degradation caused by the μth exploit in the ith asset, and q_{rh} is the neutralizing impact the rth countermeasure has on the μth exploit. Some exploits can be interrelated (e.g., designed to operate in sequence so that removing one step neutralizes the entire sequence); $\beta_{h\mu}$ is the degree of impact the hth exploit has on the μth exploit (0 if no impact and 1 if removing the former neutralizes the latter). Assets are interrelated (e.g., if assets collaborate in a mission, degrading one asset can reduce capacities of some of the others respective to that mission); α_{ji} is the degree of influence the ith asset has on the jth asset. Total degradation $F(\|\lambda_{rh}\|)$ in A is taken to be the sum of all the degrading impacts, a selection of countermeasures $\|\lambda_{rh}\|$ is sought that minimizes $F(\|\lambda_{rh}\|)$, $\lambda_{rh}=1$ if the rth countermeasure gets selected in response to the hth exploit and $\lambda_{rh}=0$ if otherwise.

The cyber warfare problem boils down to dynamic multivariable probabilistic optimization. In general, the problem poses extreme computational challenges (rendition of the problem in Eq. 1.1 contains numerous simplifying assumptions). Multiple decision aiding tools and techniques have been developed focused mainly on three aspects of the task: (1) reducing uncertainties associated with detecting and characterizing exploits [19], (2) assessing how the impact of exploits is influenced by topology and other characteristics of the network [9, 10, 24] and (3) selecting optimal countermeasures [23].

In order for the analyst to take full advantage of these methods and techniques, their outputs need to be integrated into a unified mental model enabling reversible mapping between the asset and attack compositions and giving rise to situation understanding. Per the previous discussion, mental models are concerned with (**a**) variable grouping, (**b**) value attribution and (**c**) interaction strength between the variables. That is, understanding is facilitated if intercourse between the analyst and the tools can be structured in a manner satisfying these concerns and is hampered if not. Stated differently, intelligent interface is needed allowing two-way translation between the domain data and the mental models the analyst needs to develop and

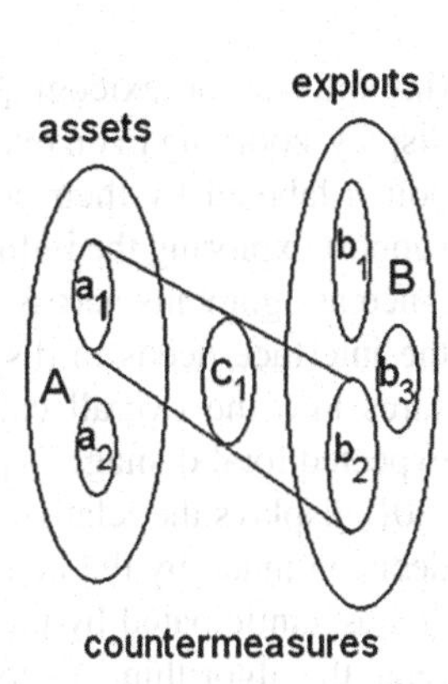

Fig. 1.10 Understanding of large and complex problems is attained by partitioning them into smaller and simpler subproblems. The subproblems are made minimally interdependent but not separate. As a result, as the conditions change, so do the subproblems (i.e., variables undergo flexible re-grouping)

maintain throughout the mission in order to comprehend the data. Some suggestions are outlined below.

1.6.1 Support Variable Grouping

This requirement implies dynamic partitioning of the overall problem into individually manageable and minimally interdependent subproblems. Subproblems involve groups of assets and groups of exploits mapped onto each other via the countermeasures, as shown in Fig. 1.10. Minimal interdependence entails the possibility of manipulating elements in one group without considering the impact these manipulations might have on the elements in other groups.

The interface needs to support partitioning of the problem stated in Eq. 1.1, as specified next.

1.6.2 Support Value Attribution and Bidirectional Mapping

Per Eq. 1.1, optimal counter measure selection is a product of interplay between value-based and probability-based characteristics of the situation. In multiobjective optimization for cyber analysis, values are equated to structural importance and reliability importance measures [23]. These measures reflect how reliability of component influences reliability of a system depending on the way the system is arranged and the position that particular component has in the arrangement [3, 17]. The measures are derived from topological and other characteristics of the network and they are not unlike fixed values attributed to chess pieces (see Sect. 1.3). Subjective value of an asset reflects the perceived degree of its relevance to the current mission, and can change as the mission progresses. The interface needs to allow entering subjective values in Eq. 1.1 (V_i, $i = 1, 2, .., S$), with subsequent determination of optimal countermeasures. Entering the values obtains a vertex-weighted asset graph which can, upon request, be partitioned into subgraphs of equal summary value (importance) or lexicographically, i.e., in the order of decreasing importance. In the display, zooming in on one of the subgraphs causes the remaining ones collapse into points labeled by their corresponding importance values. The interface needs to support exploring the influence of values variation on the graph partitioning, importance assignments across the subgraphs and countermeasures selection. Inversely, the interface needs to display the relative contribution of individual countermeasures into the overall cause of own assets protection (i.e., minimization of the expected total damage $F(||\lambda||)$. For example, an invention (called Importance Map, [20]) displays the relative impact of different malware features on the classification decision made by the neural net. Such displays can be highly informative because "maps" anticipated by the analyst can differ from those "drawn" by the algorithm (e.g., the algorithm can correctly classify the picture in Fig. 1.7 as "image of man" but count x_3 as a part of the image as opposed to a part of the surrounds).

1.6.3 Support Attributions of Interaction Strength

Subjective strength of interaction between two assets reflects their perceived importance to and the degree of collaboration in the performance of the mission. These factors are subject to change as the mission progresses. Interface needs to allow entering the subjective interaction strength values in Eq. 1.1 (the $||\alpha_{ji}||$ matrix), with subsequent determination of optimal countermeasures. Entering the strength values obtains a link-weighted asset graph that can, upon request, be partitioned into weakly coupled (minimally interdependent) subgraphs. The interface needs to support exploration of the influence of strength variation on problem partitioning, composition and the strength of coupling between the subgraphs, and the countermeasures selection.

1.6.4 Support Unification and Coordination

The interface must support the analyst in integrating the results of a, b and c above, by presenting the results in a unified and visually compelling manner. Imagery is an integral part of attaining understanding [21], the need for visual support is especially strong in complex multivariable tasks such as cyber warfare [5, 15]. A presentation format that is both logically and visually compelling and consistent with the demands of the task is suggested by the notion of cyber terrain:

> Cyber terrain is a multi-level information structure that describes cyber assets and services, and their intra- and inter-dependencies. If we look on a goal-directed activity of missions, their internal structure and time-dependent behavior then we can interpret missions as agents who "live" in a cyber terrain ([9], p. 253).

The operations of grouping, value and strength attribution are expressed naturally and in a unified fashion as changes in the terrain layout and characteristics, solutions of Eq. 1.1 (selection and allocation of countermeasures) under varying conditions (re-distribution of values) can be represented as coordinated redeployment of "agents" across the terrain [27].

1.7 Again: What is Understanding?

Understanding is not a computable function, in the following sense. Picture recognition is a computable function: Neural networks compute weighted sums of pixel values in the picture which turns out to be sufficient, in many cases, for telling apart pictures in large picture sets and telling apart components inside the pictures. In contrast, picture understanding is not reducible to a function over pixel values, for the same reasons that an understanding of chess positions does not reduce to summing up the piece values.

Patients with visual agnosia can see details of a picture but are unable to form relations between those details. When asked to describe the picture,, their verbal account is a list of details in no particular order; when prompted to reproduce the picture they draw blobs in a random juxtaposition. Patients with simultaneous visual agnosia can perceive only one object at a time. As a result, they fail in any task where an object has to be visually related to another one. The cannot, for example, place a dot in the center of a circle because they see either the circle or the tip of a pencil but not both entities at the same time [18]. In a similar way, one can have difficulty forming relations between concepts or data elements. The opposite of understanding is not misunderstanding but fragmentation.

If understanding comes not by computation, than how? Stunningly, chess computers had to reach the speeds of over 10^8 position discriminations per second in order to catch up with human players capable of, at most, a few discriminations per second. What cognitive mechanisms can be responsible for overcoming the 1: 10^8 speed disadvantage? The answer is in plain sight, and has been there for a long time. Chess masters differ from novices in their ability to form relations: Novices can perceive and operate on individual pieces while masters perceive and operate on configurations, i.e., cohesive groups of strongly interacting pieces unified in some global relationship. Masters are better than novices in recalling briefly viewed positions, which can't be explained by better visual memory: When the pieces are randomly scattered, recall accuracy is about the same in both groups [6, 7].

The crux of the matter is that forming configurations is concomitant with valuing them ("white are stronger in the center," "black have a better pawn structure," etc.). Figuratively, the chess master navigates the astronomically large combinatorial space in a capsule controlled by a joystick, which is the separation threshold x_0 in Fig. 1.8. Ability to form and attribute values to configurations is equivalent to a warp drive: By "sensing" configuration value and forming relations between configurations, one skips over vast expanses in the combinatorial space. Varying the configurations establishes value gradient pointing to promising directions for the next jump. The finding that meaningful configurations are better recalled by masters than by novices suggests strongly that configurations are perceived not as static snap-shots but as dynamic entities (that is, occupying some retrodiction-prediction interval in the variation trajectory). The gradient mechanism might explain the 15 moves look ahead accomplishment [12], which task is likely to cause overheating even in a computer doing pentaflops.

Computation (that is, executing algorithms, terms "computation" and "calculation" are used interchangeably [25]) does play a role in cognitive processes albeit a limited one. According to a chess rule of thumb, "knowing when to calculate is just as important as knowing how to calculate" ([1], p. 118). Even more to the point, "you must have a sense of when to stop (*calculations*)" ([12], p. 51). The fact is that answers to the pivotal "when" question are not computable. More precisely, no algorithm A is possible that would predict whether algorithm B working on data C is going to halt or will keep running forever [25]. Initiating and halting operations on configurations must be the work of those same mechanisms that form those configurations to begin with, and that is not computation.

A recent discovery has shed light on the nature of such mechanisms: Neuronal processes in the brain self-organize under the free energy minimization principle [8]. The discovery has established a unified framework for the analysis of cognitive and biophysical processes and opened avenues for the design of "machine understanding" (and escaping from the cul-de-sac of the computer metaphor conceptualizing brain as one big general-purpose computer or a mesh of small specialized ones). Clearly, problems having no computable solutions do not defy biology.

Relations between energy processes and cognition are discussed in [28]. Roughly, assume that classification task in Fig. 1.8 is carried out by a population of neurons where two phases are forming, A and B. The surface tension in the interphase boundary depends on the amount of free energy per unit area. The free energy reduction principle dictates reducing boundary surface thus driving the system towards more "tightly packed" subpopulations inside the phases and weaker interaction between the phases. Reaching equilibrium in some state x_0 is experienced as simultaneous emergence (perception) of two distinct entities, such as a chess configuration and its surrounds. The boundary is mobile and shifts when conditions vary which, in general, requires energy influx. To complete the picture, assume that some bulk property of the subpopulations (like temperature) is amenable to assessment (e.g., warm is good, too cold or too hot bad). That would allow "sensing" the relative values of the entities and how these values change when the conditions vary. On that proposal "attention focusing" consists in targeted energy infusion in the neuronal substrate. A number of findings in the literature appear to be consistent with this proposal, including extreme cases such as the following one: a patient suffering from simultaneous agnosia as a result of a wound in the occipital area was recovering the ability to perceive several objects simultaneously after caffeine solution was injected into the concerned area, which ability was retained for the duration of the caffeine action [18]).

One's ability to "focus attention" can decline as a result of stress, fatigue or other factors and thus become insufficient for overcoming fragmentation. Fortunately, performance can be augmented by means other than invasive interference. For example, a patient incapable of repeating a sequence of three statements began performing the task without difficulty when presented with a picture of three boxes connected by arrows and labeled "1", "2", "3" [18]. The point is that interface can help an analyst to overcome hurdles in forming coherent models of complex dynamic situations providing that the flow and organization of information presentation in the interface are consistent with the flow and organization of cognitive processes.

This proposal envisions that progress towards machine understanding will require hybrid digital/analog systems. One of the expected benefits of such hybrids is radical reduction in learning time: learning in digital computers consists in finding proper combinations of weights on the variables which is highly computationally demanding. If cognitive learning is rooted in regulated phase transition and separation in the neuronal substrate, emulating these mechanisms in the analog component will cut down computational burden in the digital one. In lieu of analog emulation, the mechanisms can be computationally approximated which will demand mathematical apparatus different from the state-of-the-art in machine learning. It is possible that, in the hybrids, many security problems will disappear and the remaining ones will be quite different from those plaguing digital computers. Switching from horse-riding to car-driving had resolved some difficulties and brought about new ones but the prospect of having one's vehicle eaten by wolves ceased to be a concern. On that happy note this article is coming to a halt.

References

1. Alburt, L., Lawrence, L. 2003. Chess Rules of Thumb. Chess Research Center, N.Y.
2. Barford, P., Dacier, M., Dietterich, T. G., Fredrikson, M., Giffin, J., Jajodia, S., Jha, S., Li, J., Liu, P., Ning, P., Ou, X., Song, D., Strater, L., Swarup, V., Tadda, G., Wang, C., Yen, J. 2010. Cyber SA: Situational awareness for cyber defense. In: S. Jajodia, P. Liu, V. Swarup, C.Wang. (Eds.) Cyber Situational Awareness. Issues and Research. Springer. pp. 3-4.
3. Birnbaum, Z.W. 1968. On the importance of different components in a multi component system. Technical Report, Washington University Seattle Lab of Statistical Research, Seattle, WA.
4. D'Amico A., Whitley, K., Tesone, D., O'Brien, B., Roth, E. 2005. Achieving cyber defense situational awareness: A cognitive task analysis of information assurance analysts. Proc. Human Factors and Ergonomics Soc. 49th Annual Meeting, pp. 229-233.
5. D'Amico A., Goodall, G.R. 2007. Visual discovery in computer network defense. IEEE Computer Graphics and Applications, 20-27.
6. de Groot, A. D. 1978. Thought and Choice in Chess. The Hague: Mouton
7. de Groot, A. D., Gobet, F. 1996. Perception and Memory in Chess: Studies in the Heuristics of the Professional eye. Assen, The Netherlands: Van Gorcum.
8. Friston, K. 2010. The free-energy principle: a unified brain theory? Nat. Rev. Neuroscience, 11, 127–138.
9. Jacobson, G. 2011. Mission cyber security situation assessment using impact dependency graphs. 14th Int. Conf. on Information Fusion, Chicago, Il, July 5-8, pp. 253-260.
10. Jajodia, S, Noel, S. 2010. Topological vulnerability analysis. In: S. Jajodia, P. Liu, V. Swarup, C. Wang. (Eds.) Cyber Situational Awareness. Issues and Research. Springer, pp. 139-154.
11. Helstrom, C.W. 1968. Statistical Theory of Signal Detection. Pergamon, N.Y.
12. Kasparov, G. 2007. How Life Imitates Chess. Bloomsbury,
13. Koestler, A. 1989. The Ghost in the Machine. Arkana Books. London, UK.

14. Kott, A. 2013. Towards fundamental science of cyber security. In: R. E. Pino (Ed.) Network Science and Security. Springer, pp. 1-13.
15. Lavigne, V., Gouin, D. 2014. Visual analytics for cyber security and intelligence. J. Defense Modeling and Simulation: Applications, Methodology, Technology, 175-199.
16. Lebiere, C., Pirolli, P., Thomson, R., Paik, J., Rutledge-Taylor, M., Staszewski, J., Anderson, J. R. 2013. A functional model of sensemaking in a neurocognitive architecture. Computational Intelligence and Neuroscience, Volume 2013, Article ID 921695, http://dx.doi.org/10.1155/2013/921695.
17. Leemis, L.M. 1995. Reliability. Probabilistic models and statistical methods. Prentice Hall, New Jersey.
18. Luria, A.R. 1973. The Working Brain. An Introduction to Neuropsychology. Basic Books, NY.
19. Mancini, L.V., Pietro, R. 2008. Intrusion Detection Systems. Springer, N.Y.
20. McLean, M. 2013. Concurrent learning algorithm and the importance map. In: R. E. Pino (Ed.) Network Science and Security. Springer, pp. 239-250.
21. Newton, N. 1996. Foundations of Understanding. John Benjamins Publishing, Amsterdam.
22. Pirolli, P., Russell, D. M. 2011. Introduction to this special issue on sensemaking. Human–Computer Interaction, 26, 1-2, 1-8
23. Roy, A., Kim, D. S., Trivedi, K.S. 2010. Cyber security analysis using attack countermeasure trees. CSIIRW'10 April 21-23, Oak Ridge, Tennessee, USA.
24. Schneier, B. 1999. Attack trees. Dr. Dobb's J. Software Tools, 24, 12, 21-29.
25. Turing, A. 1936. On computable numbers, with an application to the Entscheidungsproblem, Proc. of the London Math. Society, Series 2, 42, 230–265.
26. Tversky, A. 1977. Features of similarity. Psyc. Rev, 84, 327-352.
27. Yufik, Y.M. 2014. "I can see what I can do": Looking at complex tasks through an intelligent looking glass. In preparation.
28. Yufik, Y. M. 2013a. Understanding, consciousness and thermodynamics of cognition. Chaos, Solitons and Fractals, 55, 44-59.
29. Yufik, Y.M. 2013b. Understanding cyber warfare. In: R. E. Pino (Ed.) Network Science and Security. Springer, pp. 75-92.
30. Yufik, Y.M. 2013c. Towards a theory of understanding and mental modeling. Recent. Advances. Comp. Sci., 250-255.
31. Yufik, Y. M. 2003. Transforming data into actionable knowledge in network centric warfare. J. Battlefield Technology, 6, 1, 1-10.
32. Yufik, Y.M., Sheridan, T.B. 2002. Swiss army knife and Ockham's razor: Modeling and facilitating operator's comprehension in complex dynamic tasks. IEEE Systems, Man Cybernetics, 32, 2, 185-199.

Chapter 2
Neuromorphic Computing for Cognitive Augmentation in Cyber Defense

Robinson E. Pino and Alexander Kott

2.1 Introduction

The growth of digital content and information through the World Wide Web is increasing rapidly and more of this traffic is generated by smart mobile low size, weight, and power (SWaP) devices that are constantly sending/receiving information to/from the network for up-to-date operation. In terms of data, according to an IDC report by Gantz and Reinsel in 2012 [1], from 2005 to 2020, the digital universe will grow by a factor of 300, from 130 to 40,000 exabytes, and from now until 2020, the digital universe will about double every 2 years. The size of the digital universe in 2010 was estimated at 1,227 exabytes [1] in particular. Therefore, it can be expected that an increasing number of low SWaP devices will be implemented to offer enhanced functionality in terms of the complexity and number of services offered to users within the physically and electronically constrained form factor architecture. From a network security stand point, it will be important for the Army to ensure security and trust in the operation and functionality of smart mobile tactical devices. However, from the user's point of view, performance degradation due to security add-ons may degrade device performance during operation and during operations where speed is critical, enhanced security could degrade operational effectiveness. Therefore, it is the main goal of this effort to perform basic research in methods and techniques to provide security to mobile tactical networks while ensuring low SWaP technical requirements for operation. In this pursuit, we have considered two basic research areas that could provide a revolutionary solution to

R.E. Pino (✉)
U.S. Department of Energy, Office of Science, Washington, DC 20585, USA
e-mail: robinson.pino@science.doe.gov

A. Kott
Network Science Division, U.S. Army Research Laboratory, Adelphi, MD 20783, USA
e-mail: alexander.kott1.civ@mail.mil

R.E. Pino et al. (eds.), *Cybersecurity Systems for Human Cognition Augmentation*, Advances in Information Security 61,
DOI 10.1007/978-3-319-10374-7_2

the problem. The first technology area is memristor-based computing and the second area is artificial neural networks. It is expected that memristor-based physical computing architectures will deliver ultra-low SWaP and neural networks will enable parallelism and reconfiguration benefits. This chapter will provide a brief overview of the memristor technology and its applications within neural networks and their potential application to enabling human cognition augmentation in the Cyber-domain.

2.2 Memristor Based Technology

Interest in memristor-based technologies resurfaced in 2008 with the publication of two papers by Strukov in May 2008 [3] titled "The missing memristor found" and followed by Williams in December 2008 [4] with a paper titled "How We Found the Missing Memristor". These papers reference the pioneering early research work by Chua in 1971 [2] in which the memristor device is theorized to exist based on physical principles. The memristor is a two terminal passive device whose name is the contraction for memory resistor whose intrinsic property is to remember its previous state. We can think of the memristor as a variable resistor device whose resistance or impedance value can be changed by electrical current pulses; the electronic equivalent to the mechanical potentiometer. For example, positive current can decrease the resistance value while negative current can increase the resistance value. In addition, the memristor device exhibits electronic hysteresis which means the device current-voltage properties follow a characteristic loop behavior as shown in Fig. 2.1 [5]. From the figure, we can observe that the device behaves linearly (between −0.35 and 0.22 V approximately) and non-linearly (below −0.35 and above 0.22 V approximately). In the linear region, we can observe two states of high and low resistance. The non-linear region is the programming region that allows the device to move from a high resistance state to a low resistance state and vice-versa within the hysteresis loop as shown in Fig. 2.1 and described in more detail the published literature [5–7].

The promise memristor-based technologies offer to network security applications could be revolutionary. According to Williams [4], the potential for memristor-based applications goes towards designing new circuits that mimic aspects of the brain where neurons are implemented with transistors and synapses with memristors [4, 8]. Recently, Shevenell [10] characterized the collection performance of FPGA-based sensors compared with the current commodity server-sized sensors to be over 30 times higher collection rates with 150 times less weight and less than 4 % of the size as shown in Table 2.1. Also, Xia and his team [9] published a paper showing that memristors could improve field-programmable gate arrays (FPGA) by shrinking the processor by nearly a factor of 10 in area. Given Shevenell's and Xia's research, a memristor-based FPGA sensor could outperform the current state-of-the-art systems while reducing the form factor by 10 times. This new type of physical-memristor-based neural network computing technology could be programmed or trained to

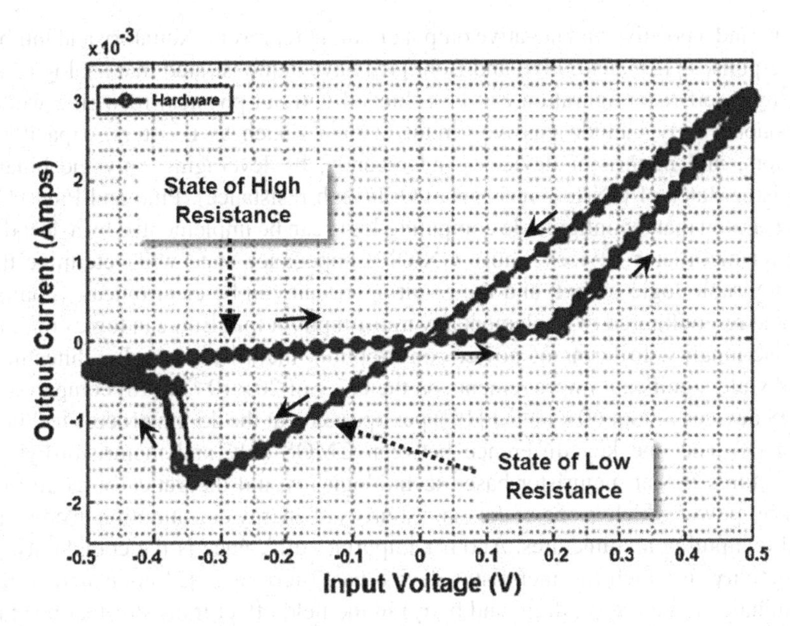

Fig. 2.1 Example of memristor device electronic hysteresis characteristic behavior, adapted from [5]

Table 2.1 Comparison of size, weight, power and throughput [10]

	Dell Power Edge R715	FPGA Xilinx ZYNQ	Raspberry PI
Weight (lbs)	49.6[a]	0.321	0.126
Size (in)	29×17×3.4[a]	8×8×1	4×3×1
Power (W)	388[a]	3–6	3.5–7
Collection Rate (Mbps)	300[b]	10,000+	66

Information obtained from obtained from [a][12] and [b][11]

perform the data analysis within a network intrusion detection system (IDS) at the sensor with enhanced performance and SWaP capabilities. Table 2.1 also compares the SWaP characteristics of the Raspberry Pi [13], a credit-card sized computer that can be used similar to a desktop PC, for spreadsheets, word-processing, games, and high-definition video.

Physical implementations of memristor-based neural networks have been already explored and characterized by Pino and his team [14–16]. For example, Pino describes an apparatus and method for dynamically and autonomously causing a circuit to reconfigure itself and produce a different output for the same input relative to the circuit's initial state; and the circuit's state remains constant until the memristor's resistance is changed, at which point the circuit's function is "reprogrammed" [14]. In addition, it has been shown how a physical memristor-based computing architecture can perform the function of neurons and synaptic connections by providing variable resistance circuits to represent interconnection strengths between neurons and a positive and negative output circuit to represent excitatory and inhibitory responses [15]. Also, Pino and Bohl [16] have demonstrated that analog resonant logic self-reconfiguration can be achieved without physical re-wiring where components only include passive circuit elements such as resistors, capacitors, inductors, and memristor devices. And recently, by leveraging only the binary memristor states of ON(low resistance)/OFF(high resistance), Pino and Pino [17] demonstrated that reconfigurable computing logic can be implemented by a decoder to select memristor devices whose ON/OFF impedance state will determine the reconfigurable logic output, and the resulting circuit can be electronically configured and re-configured to implement any multi-input/output Boolean logic computing functionality. In terms of power consumption, memristor-based architectures would yield significant improvements on the order of 20×10^{-15} J [18] compared to CMOS devices (about 60×10^{-15} J [19]) per operation at the 45 nm process technology node. The one key difference between CMOS-only and memristor-hybrid architectures is that memristor-based technologies would theoretically require no stand-by-power which is one of the main sources of energy consumption in conventional computing architectures. Another important difference is the complexity of connectivity in which the memristor device has 2-terminals [2] compared to the 4-terminals (source, gate, drain and body) in the field effect transistor that need to be properly biased for operation. While these are promising developments showing that it is possible to manufacture the technology, for very large-scale implementations of designs containing over 1,015 networked memristor devices, it will be a benefit to reduce the programming overhead of each memristor device. With this end in mind, we have developed in-house a basic neural network algorithm model called ALIEN (for Adaptive Locally Influenced Estimation Network) that simplifies simulation and implementation of memristive hardware designs while being able to demonstrate the direct applications to network security presented and discussed in the results and discussion section of the report. Appendix B shows the memristor characterization results achieve by the Army Research Laboratory in collaboration with the Air Force Research Laboratory.

2.3 Artificial Neural Networks

The memristor device's behavior of exhibiting discrete states of varying resistance is intuitively reminiscent of synaptic weights in a neural network and offers vast potential to the eventual implementation of hardware-based neuromorphic computing architectures with very modest form factor, size, weight, and power dissipation characteristics. And while neural networks are not a new construct in the field of computer science, we have developed a simple neural network learning model named ALIEN that seeks distinction by being designed from the ground to facilitate its implementation in memristive hardware. In order to accommodate the nondeterministic factors (manufacturing variability, inexact model response, etc.) currently associated with memristor production processes [20], the ALIEN learning algorithm eschews more sophisticated update schemes such as Backpropagation and instead achieves gradient descent with a rudimentary objective function (equivalent to 1 or −1 depending on the sign of the difference between a desired output and the actual output, or 0 if the actual output was correct) with a completely randomized learning rate as will be discussed in detail in the algorithm description section.

Artificial neural networks are attractive for their ability to adapt and learn to distinguish or classify patterns. In particular, the solution to a pattern separation problem is basically a problem of obtaining a criterion for distinguishing between the elements of two disjoint sets of patterns [22]. For example, if the patterns can be represented by points in a Euclidean space, one way to achieve separation is to construct a plane or a nonlinear surface such that one set of patterns lies on one side of the plane or the surface, and the other set of patterns on the other side [22]. Towards this end, multi-layered feed forward neural networks can achieve non-linear separation by applying successive linear computing operations [23, 24]. Mangasarian [22] and McCulloch and Pitts [21] have shown that both linear and nonlinear separation can be achieved by linear programming. In generic terms, neural networks offer the potential to generalize a problem by learning to detect patterns from data using a parallel distributed computing architecture, and the key to program, train, or teach a neural network to perform such a function is the learning algorithm. In the published scientific literature, there are several learning algorithms that can be used to train neural networks; however, from a pragmatic implementation perspective, the Backpropagation algorithm has been used for several decades as the de-facto learning algorithm to train and prototype multi-layered neural network-based systems [23, 24]. Recent neural network research and development employs advances in computational methods and mathematical techniques to achieve fast convergence rates and accuracy; however, it is the goal of our in-house research to simplify this approach to map a low mathematical complexity neural network learning algorithm for applications with physical memristor-based neural networks that will minimize circuit overhead in design and manufacturing while preserving the adaptive properties of artificial neural networks.

2.4 ALIEN Learning Algorithm

The main purpose of the ALIEN learning algorithm is to minimize the number and complexity of computations required to train a neural network. Figure 2.2 shows a two input and one output neural network with a single hidden layer, and this will be used to explain the ALIEN learning algorithm.

The feed forward operation of each neuron, Nn, is that of a threshold summing node in which whenever the summed product of the input times the weight, Wn, over all inputs m is greater than a given threshold value, Vth, the output of the neuron will be 1 otherwise −1 as shown in Eq. 2.1. Therefore, applying this feed forward operation from layer to layer will result in an output. From this formalism,

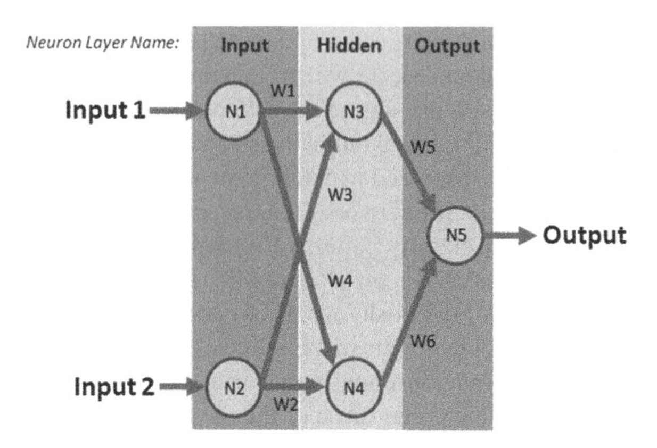

Fig. 2.2 Feed forward neural network configuration with two inputs, two hidden and one output (five neurons) and six weights or synaptic connections

we can see that the only controlling parameter to change the neuron's output is the weight element, Wn.

$$\text{Neuron Output}, N_n = \begin{cases} 1 & \text{if} \sum_{n=1}^{m} W_n \times \text{Input}_n \geq V_{th} \\ -1 & \text{otherwise} \end{cases} \tag{2.1}$$

The algorithmic operation to update each weight element, Wn, is a follows:

Compute output neuron direction (OND) by taking the sign of the subtraction between the desired output and actual output of the network. The output neuron direction is the polarity of the output of the neuron which needs to be corrected up or down (plus or minus).

$$\text{OND} = \text{Desired Output} - \text{Actual Output} \tag{2.2}$$

Compute hidden layer neuron direction (HLND) by taking the sign of the multiplication between the output neuron direction and connecting weight element. For more than one output to which the hidden neuron is connected, compute the sign of the total sum for all output neuron directions times the weight connecting the hidden neuron to any output neuron.

$$\text{HLND} = \sum_{n=1}^{m} \text{OND}_n \times \text{Weight}_n \tag{2.3}$$

Compute change value to output weights (ΔOW), which connect the hidden layer to the output layer neurons, by multiplying: output neuron direction, learning rate, random value and hidden layer neuron output (HLNO). The learning rate is any positive number normally ranging from 10 to 0.00001, zero means not learning and smaller value means slower convergence. The random value is any value between 0 and 1.

$$\Delta\text{OWn} = \text{ONDn} \times \text{LR} \times \text{RAND} \times \text{HLNOn} \tag{2.4}$$

Compute change value to hidden weights (ΔHW), which connect the inputs to the hidden layer neurons, by multiplying: hidden neuron direction, learning rate, random value and input to the hidden neuron or input layer neuron output (ILNO).

$$\Delta \mathrm{HWn} = \mathrm{HNDn} \times \mathrm{LR} \times \mathrm{RAND} \times \mathrm{ILNOn} \quad (2.5)$$

Update weights by adding the weight change value to each synaptic weight element.

$$\mathrm{Wn} = \mathrm{Wn} + \Delta \mathrm{Wn} \quad (2.6)$$

During our discussion on learning and training of the neural network two important terms will be mentioned iteration and epoch. Training or learning iteration is the process of applying the ALIEN learning algorithm to update the synaptic weights in the neural network for the entire training dataset once. The epoch is the process of generating new random synaptic weights for all synaptic weights in the neural network.

In order to validate ALIEN's feasibility as a machine learning algorithm, we sought to demonstrate its ability to correctly learn the two-bit Boolean exclusive OR (XOR) function and 8-bit odd Boolean parity. These results are shown in Appendix A. Then, we implemented ALIEN to solve two network security challenges: First to classify malicious traffic from network flows; and second to classify A or MX DNS record requests based only on the network packet data. For context, let's take a look a how neural networks have been applied in network security applications.

2.4.1 *Learning Algorithm Characterization*

2.4.1.1 XOR Function

The initial experiment to demonstrate the capability of ALIEN was to learn the function of the XOR Boolean function as shown in Fig. 2.3. The importance of the XOR function is that it is a simple two-input and one-output non-linear classification problem. Therefore, it is important to demonstrate the neural network can perform this basic non-linear classification problem by learning to perform the XOR function perfectly.

The results of the trained ALIEN neural network are shown in Fig. 2.4. The figure displays the percentage of successful convergence and not successful convergence versus learning rate. The maximum number of iterations allowed during the experiment was 10,000 iterations. Whenever the training process took in excess of 10,000 iterations, the experiment was determined to be not successful. From the Figure, we can observe that for low learning rates, below 0.001, the percentage of successful convergence is below 50 % and becomes lower with lower rates. Figure 2.5, shows the average and median number of iterations required for the network to converge to a perfect solution. The total number of epochs was 50 with

Fig. 2.3 The complete XOR function input and output data set

Input A	Input B	Output C
0	0	0
0	1	1
1	0	1
1	1	0

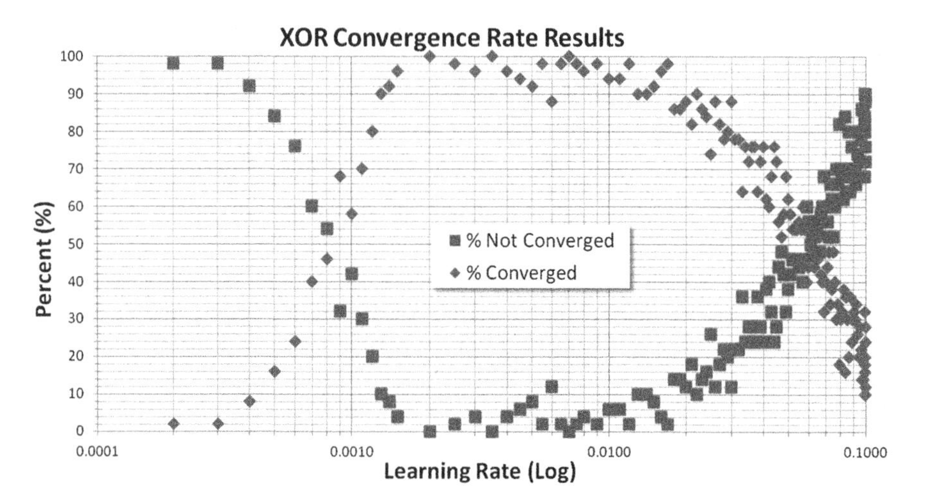

Fig. 2.4 XOR function percentage convergence versus learning rate (in log x-axis scale)

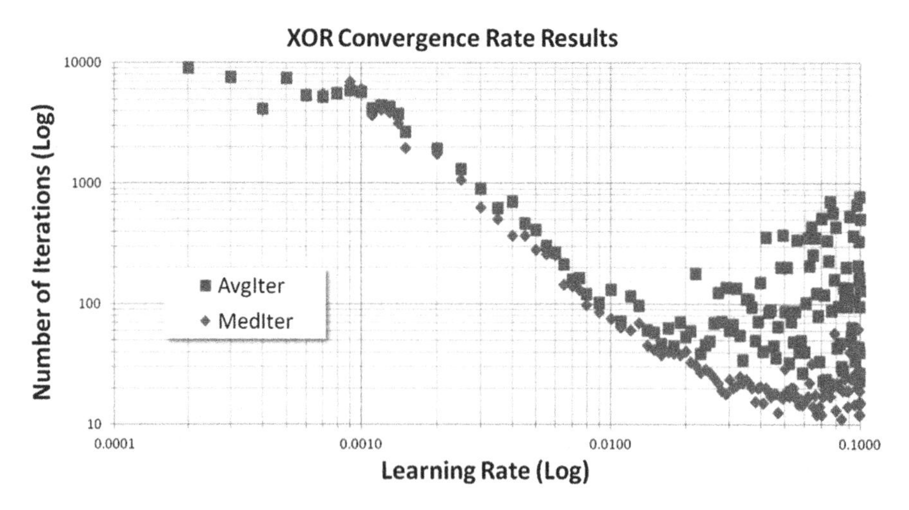

Fig. 2.5 XOR function number of average (*square*) and median (*diamond*) iterations required for a successful convergence versus learning rate (in log x-axis scale)

four hidden neurons and one output neuron. From the Figure, we can observe that as the learning rate was increased, the median number of iterations required to achieve a perfect solution were lower while the average number of iterations became noisier. From these results, we can conclude that for small learning rates, the number of iterations required to achieve the perfect solutions increases as the synapses converge slowly toward the solution. Conversely, as the learning rate becomes large, the median number of steps to achieve a perfect solution is reduced; however, the lower number of iterations required comes at the cost of increased likelihood of failing to reach a comprehensive solution within a training epoch. With a large maximum learning rate, the network may continually find itself on different sides of the ideal configuration because its adjustments lack the granularity to settle down in the global error minimum.

Figure 2.6 shows the percent successful and unsuccessful as a function of learning rate in a linear scale. From the figure, it is easy to point to the region yielding a 50 % success rate of converging to a perfect solution between learning rates from 0.05 to 0.07 approximately. In addition, we can observe that as the learning rate becomes larger, the distribution of successful and not successful convergence grows and widens as the learning rate grows. Figure 2.7 shows the average and median number of iterations required to achieve successful convergence. In particular, it is interesting to note that from an approximate learning rate between 0.02 and 0.1, the median iteration rate stays consistently below 100; while on the other hand, the average number of iterations required grows to an upper bound of close to 1,000 iterations. In general, from this experiment, we can determine the most suitable parameters to achieve reliable convergence when performing the two-input nonlinear classification problem of the XOR Boolean function. From Fig. 2.6, we can

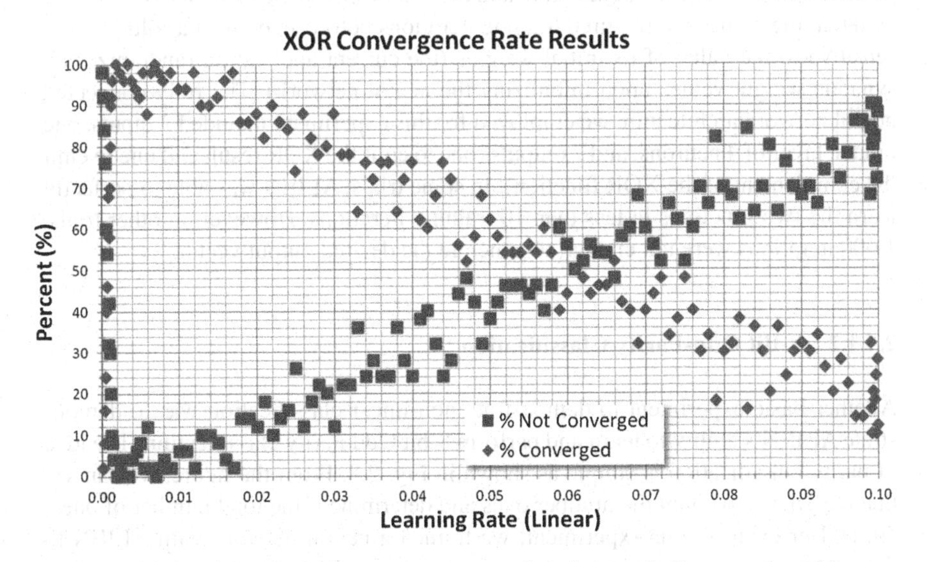

Fig. 2.6 XOR function percentage convergence versus learning rate (in linear x-axis scale)

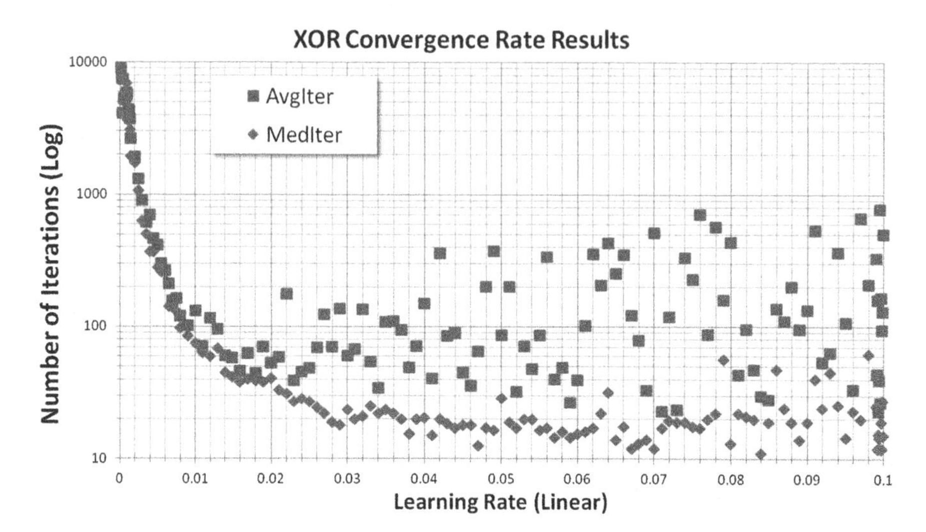

Fig. 2.7 XOR function number of average (*square*) and median (*diamond*) iterations required for a successful convergence versus learning rate (in linear x-axis scale)

see that at a learning rate of 0.02, we could achieve a 90 % convergence rate that would require approximately 100 iterations to achieve perfect learning.

From our experimental results, we found that the ALIEN model is very capable of perfectly learning the 2-bit Boolean XOR function shown in Fig. 2.3 as well as the complement of the function; that is, the inverse of the output. One interesting aspect however was to characterize and explore the convergence space of ALIEN given its simple learning architecture and design. From our experiments, we discovered that the number of required training iterations before arriving at a solution was sensitive to the value of its initial weights (the current application randomizes all synaptic weight values upon initializing the neural network). The fully-connected neural network architecture we employed for this experiment included 2 inputs, one hidden layer of 4 neurons, and 2 outputs (one to emit the XOR result and one to emit the complement of the XOR function). In some trials, ALIEN was able to perfectly learn the two functions in less than 10 training iterations. However, in other trials, 1,000 iterations were not enough to generate the desired output vector.

2.4.1.2 8-Bit Odd-Parity Classification

Another basic experiment to demonstrate non-linear classification was to demonstrate ALIEN's ability to learn and perform 8-bit Odd Parity classification. The idea of parity classification is simple as shown in Fig. 2.8. From the figure, we can see that the goal is to count the number ones and determine if the total number of one's are odd or even. In our experiment, we trained a neural network with ALIEN to determine if the input vector presented was odd or not. Therefore, we constructed an

Bit-1	Bit-2	Bit-3	Bit-4	Bit-5	Bit-6	Bit-7	Bit-8	Parity
0	0	0	0	0	0	0	0	Even
0	0	0	0	0	0	0	1	Odd
0	1	0	1	0	1	0	1	Even
0	0	1	0	1	0	0	1	Odd

Fig. 2.8 Examples of 8-bit even and odd parity

ALIEN neural network containing 8 inputs, one hidden layer containing 256 neurons, and one output. The result was that ALIEN was capable of reliably determining 8-bit odd parity with perfect accuracy.

The network was able to perfectly emit correct odd bit parity within just 478 training iterations in some trials while in others up to 1,675 iterations were required. As with the 2-bit XOR implementation, this variance suggests that required iterations are highly sensitive to the random synaptic weights at the beginning of the training epoch as it is expected given the random design nature of the initialization process. Therefore, we performed another set of characterization experiments to determine the behavior and suitability of the ALIEN neural network learning algorithm to perfectly learn the 8-bit odd parity function. Similarly to our previous experiment, the maximum number of iterations was set to 10,000, and the total number of epochs performed was seven given the limitations of our simulation computing platform.

Figure 2.9 shows the percent rate of successful and not successful convergence to learning the function perfectly as a function of learning rate (x-axis in log scale). From the figure, we can observe that for low learning rates the successful convergence rate is at or very close to 100 % consistently. Figure 2.10 displays the average and median number of iterations required to achieve perfect convergence. From the figure, it is shown that for learning rates between 0.00001 and 0.0001, the average and median number of iteration required to achieve perfect convergence is lower than 3,000. In particular, we can observe that the number of iterations reduces as the learning rate increases.

Figure 2.11 shows the percentage of successful and not successful convergence versus learning rate in linear scale (x-axis). From the figure, we can observe that the region that yielded a 50 % convergence rate is between 0.0006 and 0.0008. In particular, the figure also shows how quickly the successful convergence rate drops exponentially towards zero as the learning rate reaches 0.001. Figure 2.12 displays the average and median number of iterations required to achieve perfect convergence versus learning rate in linear scale (x-axis). From the figure, it is clear that the number of iteration jumps to very close to 10,000 required iterations as the learning rate increases from 0.0002 to 0.0008. In general, the optimum parameters for ALIEN to train a neural network to perform 8-bit odd parity can be determined to be at learning rate of about 0.0001 that would require an estimated 1,000 iterations .

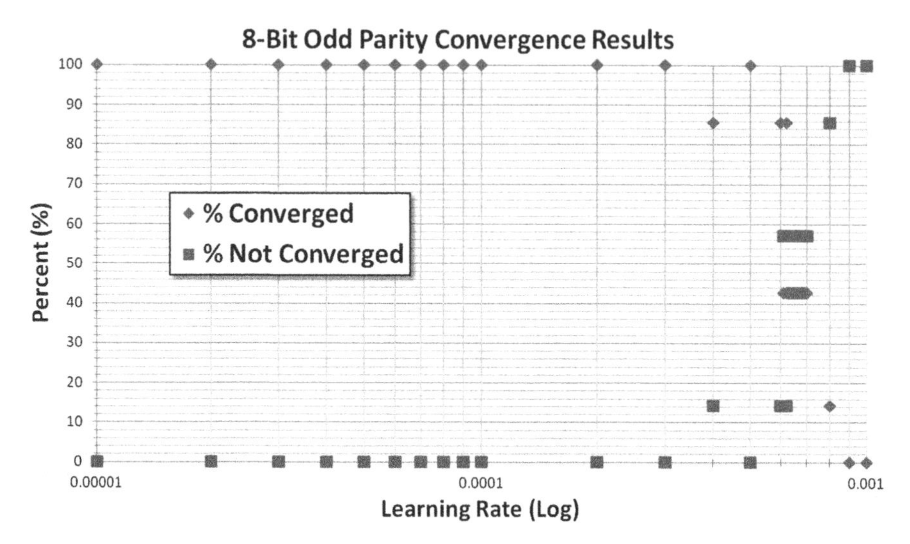

Fig. 2.9 8-Bit ODD parity function percentage convergence versus learning rate (in log x-axis scale)

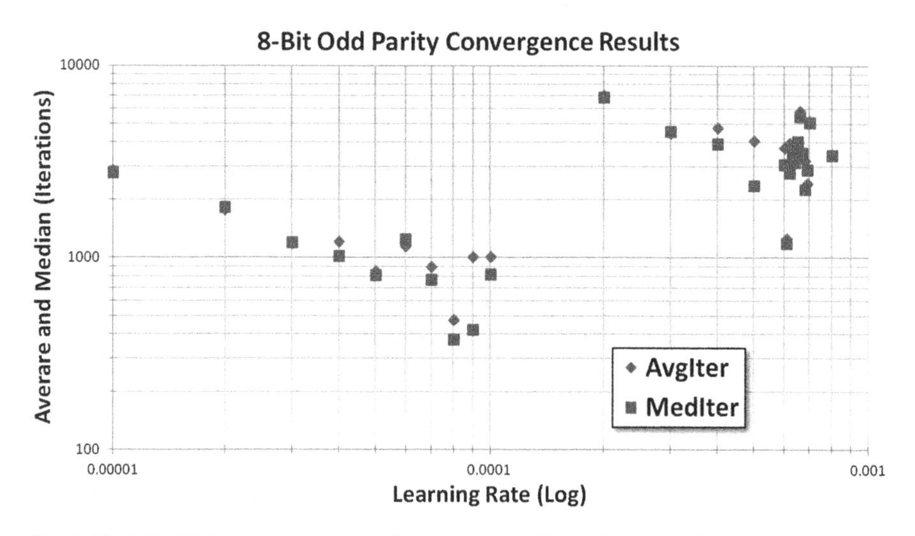

Fig. 2.10 8-Bit ODD parity function number of average (*diamond*) and median (*square*) iterations required for a successful convergence versus learning rate (in log x-axis scale)

Finally, we characterized the performance of the neural network to classify in the event that it was trained with missing or partial information. Therefore, we designed an experiment with the 8-bit odd parity data set to characterize the response. The experiment performed consisted of 70 epochs, a maximum of 2,000 iterations, at a fixed learning rate of 0.0001. The neural network configuration was the same as described previously for the 8-bit odd parity characterization experiment above.

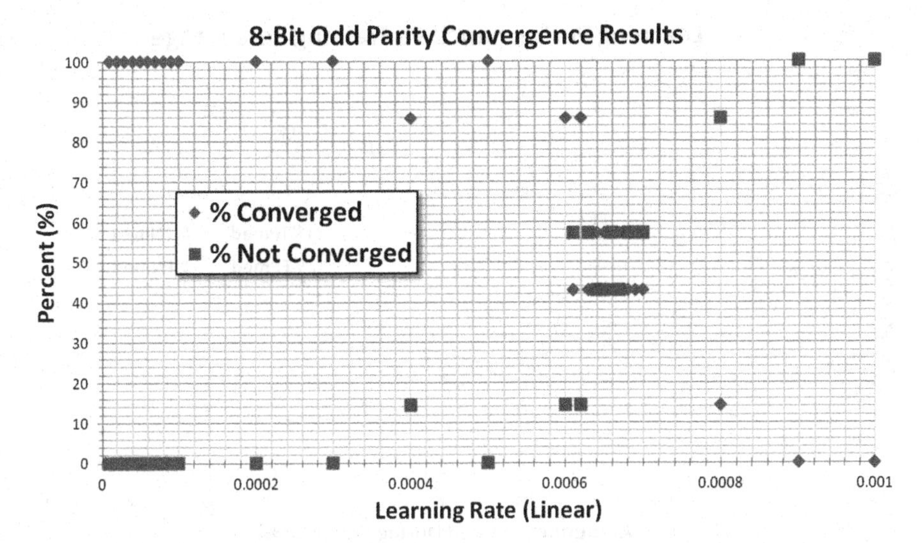

Fig. 2.11 8-Bit ODD parity function percentage convergence versus learning rate (in linear x-axis scale)

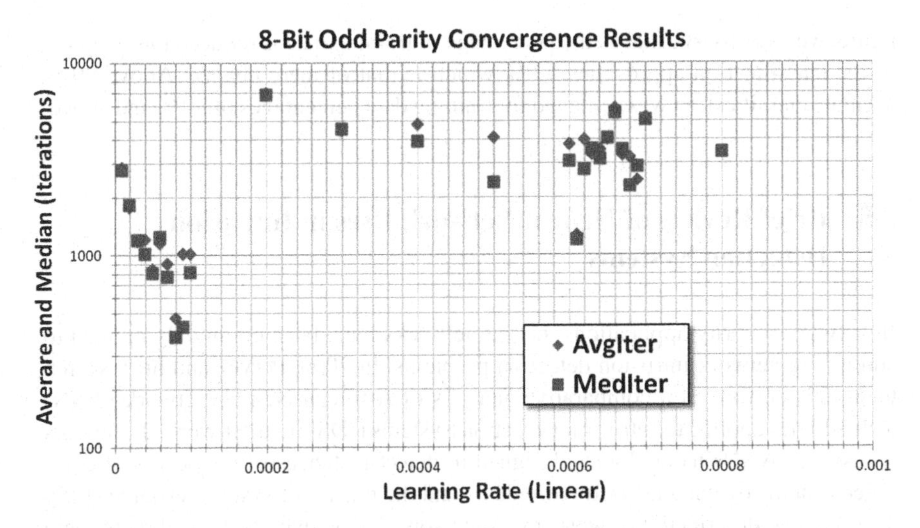

Fig. 2.12 8-Bit ODD parity function number of average (*diamond*) and median (*square*) iterations required for a successful convergence versus learning rate (in log x-axis scale)

In this case, the input training dataset was randomly selected during each epoch to contain 100, 95, 90, 80, 70, 60, 50, 40, 30, 20, 10, and 5 % of the complete dataset. For example, the last case scenario contained only 5 % of the original 256 record data set. The test phase contained the complete dataset and the True Positive (TP), True Negative (TN), False Positive (FP), and False Negative (FN) rates were measured for each record. Figure 2.13 shows the results of this experiment. From the

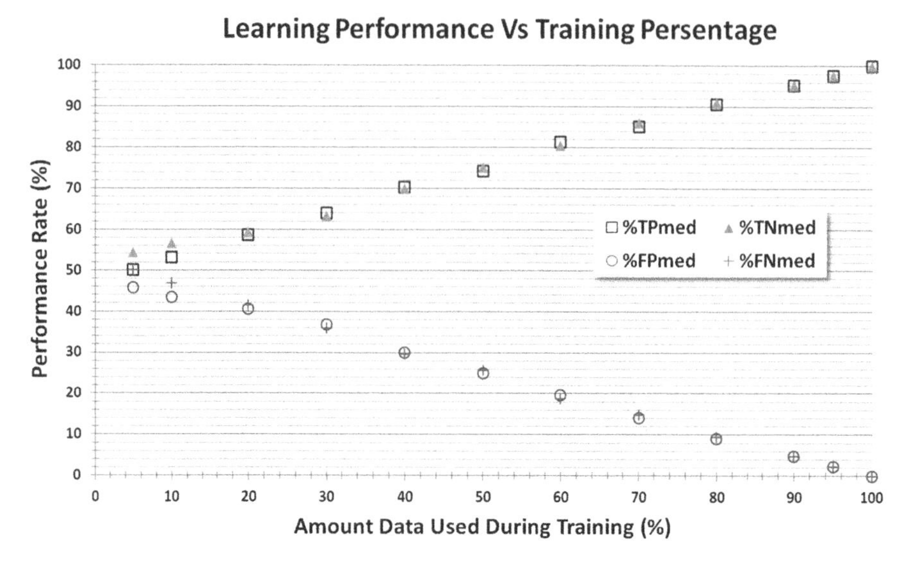

Fig. 2.13 Neural network performance as function of percent data used during training

figure, we can observe that as the percentage of training data is reduced the accuracy of the classification approaches 50 % which is that of mere chance. Above 90 % accurate classification was achieved for training data percentages of 80 % and above.

2.5 Brief Review of Neural Network Uses in Intrusion Detection Systems

In network security applications, neural networks have been commonly applied to addressing network intrusion detection problems. In 2006, Pervez and his research team [25] performed a comparative analysis of Artificial Neural Network (ANN) technologies applied to Intrusion Detection Systems (IDS). In their work, the authors focused on ANNs technologies designed to detect instances of the access of computer systems by unauthorized individuals and the misuse of system resources [25]. The authors described the work by Anderson [26] in which three threats were described which could be identified from audit data as: (1) External Penetrations—Unauthorized users of the system who gain access to the system; (2) Internal Penetrations—Authorized system users who utilize the system in an unauthorized manner; and (3) Misfeasors—Authorized user who misuse their access privileges. These threats represent areas in which a neural network could potentially be trained to identify patterns of malicious cyber behavior. Denning in 1987 provided some of the early work on intrusion detection systems which relied on identifying abnormal patterns of system usage [27]. Denning developed a model of a real-time intrusion-detection expert system capable of detecting break-ins, penetrations, and other forms of computer abuse. The model is based on the hypothesis that security violations can be detected by monitoring a system's audit records for abnormal patterns of system usage. The model includes profiles for representing the behavior of subjects in terms of metrics and statistics, and rules for acquiring knowledge about this behavior from audit records and for detecting anomalous behavior. The model is independent of any particular system, application environment, system vulnerability, or type of intrusion, thereby providing a framework for a general-purpose intrusion detection expert system [27]. Overall, the model provides a framework for IDS research and development. For misuse detection, Cannady in 1998 [28] presented an approach utilizing the analytical strengths of neural networks and provided results indicating a data correlation rate of 97.56 %. In the experiments, the training of the neural network was performed using the Backpropagation algorithm for 10,000 iterations of selected training data for which 1,000 records were used for testing out of a total of 9,462 records available in their dataset. Their results demonstrated upwards of over 90 % correlation for various types of attacks that included SYNFlood, SATAN, and ISS Scan attack tests [28]. Their results show the potential in using feed forward neural networks to identify specific attacks with a high degree of accuracy. The network traffic fields employed during training included:

- Protocol ID—The protocol associated with the event, (TCP=0, UDP=1, ICMP=2, and Unknown=3).
- Source Port—The port number of the source.
- Destination Port—The port number of the destination.
- Source Address—The IP address of the source.
- Destination Address—The IP address of the destination.
- ICMP Type—The type of the ICMP packet (Echo Request or Null).
- ICMP Code—The code field from the ICMP packet (None or Null).
- Raw Data Length—The length of the data in the packet.
- Raw Data—The data portion of the packet.

Another approach by Vollmer and Manic in 2009 involved researching computationally efficient implementations of neural networks in anomaly-based intrusion detection systems [29]. Their results maintained output accuracy while reducing by 70 % the amount of memory required to run the system. The neural network consisted of three fully connected layers and was trained and tested using ICMP rules and test data. One of the main advantages to employing neural networks in performing network security lies in the ability to store the various rule attack knowledge pattern vectors within the synaptic weights in between the neurons and layers [29] (graphically shown in Fig. 2.2) resulting in enhanced memory utilization and performance improvement.

2.6 Experimental Results and Discussion

It is clear artificial neural networks have been applied to various IDS challenges. However, it is difficult to compare any of the reported results given that most datasets employed are not publically available. Another point is whether those datasets

Table 2.2 Characteristics of the KDD dataset employed for training and characterization

Dataset Label	Total Attack	Total Normal	Total Records
Test data, corrected labels (for training)	250,436	60,593	311,029
Full data set (for test)	3,925,651	972,780	4,898,431

are relevant to DoD or military network operations. Therefore, we will compare our results on the classification of malicious flow network traffic to the results from a paper by Kayacik and team [30] that is based on a hierarchy of Self-Organizing Feature Maps. The goal in their work was to classify the KDD benchmark tagged dataset from the International Knowledge Discovery and Data Mining Tools Competition [31]. This is the data set used for The Third International Knowledge Discovery and Data Mining Tools Competition, which was held in conjunction with KDD-99 The Fifth International Conference on Knowledge Discovery and Data Mining. The competition task was to build a network intrusion detector, a predictive model capable of distinguishing between "bad" connections, called intrusions or attacks, and "good" normal connections. This dataset contains a standard set of information to be audited, which includes a wide variety of intrusions simulated in a military network environment [31]. Overall, though this dataset is old, this is the most quoted dataset in the open published literature to the best of our knowledge. One interesting point of the work by Kayacik et al. is that they attempted to perform the classification using only 6 fields of the possible 41 features available in the dataset [30]. This goal resonates with our own in-house research goal of minimizing the form factor during IDS operations, which in this case was accomplished by reducing the total number of inputs to perform the classification. From their results, the detection rate on the test set varied between 89 and 99.7 % depending on the dataset partition employed [30].

To perform the neural network training with ALIEN, both the training and test data was used described in Table 2.2. The training dataset is composed of over 311,000 flow records, and the characterization, full data set contains close to 5 million flow records. For characterizing and measuring the performance of the ALIEN trained neural network only six fields were used and are described in Table 2.3. The description of the reminder of the 35 fields can be found on the KDD repository website [31].

During our experiments, in order to further minimize the size of our classifying neural network, we created a single layer perceptron neural network architecture as shown in Fig. 2.14. From the figure, we can observe that only six input neurons, six synaptic weights, and one output neuron were implemented. In addition, Table 2.4 shows an example training input vector from the KDD dataset that is mapped to the actual training input employed to the neural network. The example in Table 2.4 shows a traffic flow record corresponding to a warezmaster attack. In addition, the complete vector value mapping for the text fields 2, 3, and 4 in the KDD dataset and the desired neural network output (the last two columns on the right, traffic type label and neuron output) are shown in Table 2.5. In particular, the order input value from Table 2.5 corresponds to the input neural network value for fields 2, 3, and 4

Table 2.3 KDD fields used during training and performance characterization

• Field 1: *duration*, length (number of seconds) of the connection
• Field 2: *protocol_type*, type of the protocol, e.g. tcp, udp, etc.
• Field 3: *service*, on the destination, such as FTP, HTTP, Telnet, etc.
• Field 4: *flag*, normal or error status of the connection
• Field 5: *src_bytes*, number of data bytes from source to destination
• Field 6: *dst_bytes*, number of data bytes from destination to source

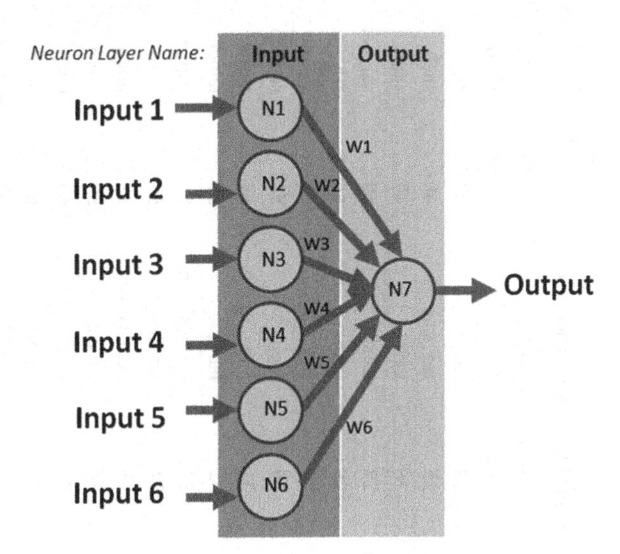

Fig. 2.14 Neural network experimental setup with a total of six input neurons, one output neuron, and six synaptic connections

Table 2.4 Example of KDD fields used during training conversion to neural network input

KDD name	Duration	Protocol_ type	Service	Flag	Src_ bytes	Dst_ bytes	Traffic type
Input	Input 1	Input 2	Input 3	Input 4	Input 5	Input 6	Output
Actual	280	tcp	ftp_data	SF	283618	0	warezmaster
Formatted	280	2	17	10	283618	0	−1

of the KDD dataset. Also, the last two columns on the right of Table 2.5 correspond to the desired neural network output for the traffic_type label field where for all possible outputs only the normal output is set to 1, all the other possible traffic_type labels are set to -1 which correspond to the malicious network traffic. Given that the KDD training and characterization datasets are relatively large, 47.3 and 742.6 MB respectively, and given the limited computing performance of the desktop machine use to run the simulations, the training experiment was run for a maximum of three

Table 2.5 Input mapping where the order value represents the input to the neural network

Order	Protocol Type	Service	Flag	Order	Service	Order	Service	Traffic Type	Neural
Input	Field 2	Field 3	Field 4	Input	Field 3	Input	Field 3	Label	Output
1	icmp	auth	OTH	**24**	IRC	**47**	rje	back	–1
2	tcp	bgp	REJ	**25**	iso_tsap	**48**	shell	buffer_overflow	–1
3	udp	courier	RSTO	**26**	klogin	**49**	smtp	ftp_write	–1
4		csnet_ns	RSTOS0	**27**	kshell	**50**	sql_net	guess_passwd	–1
5		ctf	RSTR	**28**	ldap	**51**	ssh	imap	–1
6		daytime	S0	**29**	link	**52**	sunrpc	ipsweep	–1
7		discard	S1	**30**	login	**53**	supdup	land	–1
8		domain	S2	**31**	mtp	**54**	systat	loadmodule	–1
9		domain_u	S3	**32**	name	**55**	telnet	multihop	–1
10		echo	SF	**33**	netbios_dgm	**56**	tftp_u	neptune	–1
11		eco_i	SH	**34**	netbios_ns	**57**	time	nmap	–1
12		ecr_i		**35**	netbios_ssn	**58**	tim_i	normal	**1**
13		efs		**36**	netstat	**59**	urp_i	perl	–1
14		ex ec		**37**	nnsp	**60**	uucp	phf	–1
15		finger		**38**	nntp	**61**	uucp_path	pod	–1
16		ftp		**39**	ntp_u	**62**	vmnet	portsweep	–1
17		ftp_data		**40**	other	**63**	whois	rootkit	–1
18		gopher		**41**	pm_dump	**64**	X11	satan	–1
19		hostnames		**42**	pop_2	**65**	Z39_50	smurf	–1
20		http		**43**	pop_3			spy	–1
21		http_443		**44**	printer			teardrop	–1
22		icmp		**45**	private			warezclient	–1
23		imap4		**46**	remote_job			warezmaster.	–1

to five iterations and one epoch each time. Typical results after training the neural network with ALIEN for 5 iterations and 1 epoch showed a detection rate between 96.07 and 98.46 %. These results are very encouraging for such simple neural network architecture.

The second classification experiment performed was to train a feed forward neural network using ALIEN to perform classification between DNS A and MX network traffic record requests. An MX record or mail exchanger record is a type of record in the Domain Name System (DNS) that specifies a mail server that is responsible for accepting email exchanges for the particular domain of the recipient. The A type address record returns the 32-bit IPv4 address, and it is commonly used to map hostnames to the IP address of the host.

The process to capture an MX or A record can be accomplish by using tcpdump, which is commonly used open source command line software for packet capture and analysis [33]. The specific command that is used to capture a single network packet can be written as:

```
$ sudo tcpdump -i any -nnvXA -c 1 dst port 53 > filename
```

The commands to request an MX or A packet from the network can be written, for example, as:

```
$ dig www.google.com mx
$ dig www.google.com a
```

In the command lines above, A and MX DNS records were requested from the www.google.com domain name and the results are shown in the figures below for each DNS record request respectively.

Figure 2.15 displays a typical tcpdump output example of an MX network packet as was used to perform our training. Figure 2.16 displays a typical tcpdump output example of an A network packet that was used to perform our training and classification experiment. Per RFC 1035 [32], the location that tells us whether the record represents an A or MX DNS request follows the QNAME byte sequence describing the queried domain name (www.google.com in the above example). For our specific training experiment, the packets where truncated to a length of 125 bytes containing only the hexadecimal information as shown in Figs. 2.17 and 2.18 for each of the MX and A DNS requests. For any truncated packet that was shorter than 125 bytes, null bytes (0×00) were used to pad the missing length to 125 bytes as displayed in the figures. In addition, each packet was labeled as indicated in Figs. 2.17 and 2.18 to identify the request as either an MX lookup or an A lookup.

Finally, each byte was converted to unsigned integer in the inclusive range between 0 and 255. The request type label of each packet was then represented as −1 for MX or 1 for A as shown in Figs. 2.19 and 2.20. The training records as actually submitted to the neural network are exemplified by Figs. 2.19 and 2.20.

During the MX and A DNS record classification experiment, 5,000 A and 32 MX DNS record request packets were captured employing the tcpdump software application respectively. In summary, ALIEN was able to successfully train a 63 input, 10 hidden, and 1 output neuron network configuration at a learning rate of 0.1 within

```
23:37:07.076229 IP (tos 0x0, ttl 64, id 31915, offset 0, flags [none], proto
UDP (17), length 71)
     127.0.0.1.51324 > 127.0.1.1.53: 56852+ [1au] MX? www.google.com. (43)
      0x0000:  4500 0047 7cab 0000 4011 fef8 7f00 0001  E..G|...@.......
      0x0010:  7f00 0101 c87c 0035 0033 ff46 de14 0120  .....|.5.3.F....
      0x0020:  0001 0000 0000 0001 0377 7777 0667 6f6f  .........www.goo
      0x0030:  676c 6503 636f 6d00 000f 0001 0000 2910  gle.com.......).
      0x0040:  0000 0000 0000 00                        .......
```

Fig. 2.15 MX DNS network packet captured with tcpdump, the byte containing the record type is *underlined* and in *boldface*

```
23:34:53.162407 IP (tos 0x0, ttl 64, id 31914, offset 0, flags [none], proto
UDP (17), length 71)
     127.0.0.1.55133 > 127.0.1.1.53: 53846+ [1au] A? www.google.com. (43)
      0x0000:  4500 0047 7caa 0000 4011 fef9 7f00 0001  E..G|...@.......
      0x0010:  7f00 0101 d75d 0035 0033 ff46 d256 0120  .....].5.3.F.V..
      0x0020:  0001 0000 0000 0001 0377 7777 0667 6f6f  .........www.goo
      0x0030:  676c 6503 636f 6d00 0001 0001 0000 2910  gle.com.......).
      0x0040:  0000 0000 0000 00                        .......
```

Fig. 2.16 A DNS network packet captured with tcpdump, the byte containing the record type is *underlined* and in *boldface*

```
450000477cab00004011fef87f0000017f000101c8
7c00350033ff46de140120000100000000000010377
777706676f6f676c6503636f6d00000f0001000029 MX?
```

Fig. 2.17 Truncated hex representation of an MX DNS lookup, the byte containing the record type is *underlined* and in *boldface*

```
450000477caa00004011fef97f0000017f000101d7
5d00350033ff46d2560120000100000000000010377
777706676f6f676c6503636f6d0000010001000029 A?
```

Fig. 2.18 Truncated hex representation of an A DNS lookup, the byte containing the record type is *underlined* and in *boldface*

```
69,0,0,71,124,171,0,0,64,17,254,248,127,0,0,1,127,0,1,1,200,
124,0,53,0,51,255,70,222,20,1,32,0,1,0,0,0,0,0,1,3,119,119,
119,6,103,111,111,103,108,101,3,99,111,109,0,0,15,0,1,0,0,0,41,1
```

Fig. 2.19 Neural network input for MX network packet, the byte containing the record type is *underlined* and in *boldface*

```
69,0,0,71,124,170,0,0,64,17,254,249,127,0,0,1,127,0,1,1,215,
93,0,53,0,51,255,70,210,86,1,32,0,1,0,0,0,0,0,1,3,119,119,
119,6,103,111,111,103,108,101,3,99,111,109,0,0,1,0,1,0,0,0,41,-1
```

Fig. 2.20 Neural network input for A network packet, the byte containing the record type is *underlined* and in *boldface*

approximately 30 iterations. ALIEN's classification was perfect in all but two MX records where the telltale byte identifying the lookup type was outside the truncation window. This means that for variable domain names, neural networks are not suitable for classification as the number of inputs must remain the same. A possible solution is to have very long inputs up to the maximum possible number of bytes; however, the performance of the neural network would be decrease as the computing resources would increase exponentially. Still the overall classification rate was 99.96 % with only the two MX packets not achieving proper classification as the required information was missing from the truncated record.

After having achieved success with ALIEN and learning its limitations with variable length domain names, we designed an experiment in which to break down the problem intelligently to allow feature extraction to clearly highlight the problem in question. For example, in our DNS A and MX record request classification problem, the challenge to solve is simply to find and correctly classify two specific requests. Therefore, the problem should be easy for a neural network to solve as the choices are only limited two possible solutions, A or MX. In the updated approach to feature extraction, the neural network inputs are the respective MD5 digests of the DNS question QNAME, QTYPE, and QCLASS fields normalized to a decimal representation between −1.0 and 1.0. Using a one-way hashing function digest to represent the QNAME, QTYPE, and QCLASS fields allows the classifier to accept these fields as distinct features free from dependency upon the length or byte-offset position of each field. Since there is no cryptographic strength requirement for this particular hashing application, MD5 was chosen as an expedient option because an implementation already exists within the "hashlib" package of the standard CPython 2.7.3 distribution. Figs. 2.21 and 2.22 describe how the question section of the DNS request packet was divided into three distinct inputs that cover the domain name (www.google.com, used as an example), query type (A and MX), and the query class of the lookup being performed.

The experimental results of the training performed using the new methodology is displayed in Figs. 2.23 and 2.24.

Question section of DNS A lookup	\x03www\x06google\x03com\x00\x00\x01\x00\x01
Input vector divided into three parts	\x03www\x06google\x03com\x00 , \x00\x01 , \x00\x01
Hashed inputs to Neural Network	-0.521159634814 , 0.0901948225429 , 0.0901948225429 , -1

Fig. 2.21 DNS A network packet and conversion to neural network input

Question section of DNS MX lookup	\x03www\x06google\x03com\x00\x00\x0f\x00\x01
Packet divided into three inputs	\x03www\x06google\x03com\x00 , \x00\x0f , \x00\x01
Hashed inputs to Neural Network	-0.521159634814 , -0.434861249159 , 0.0901948225429 , 1

Fig. 2.22 DNS MX network packet and conversion to neural network input

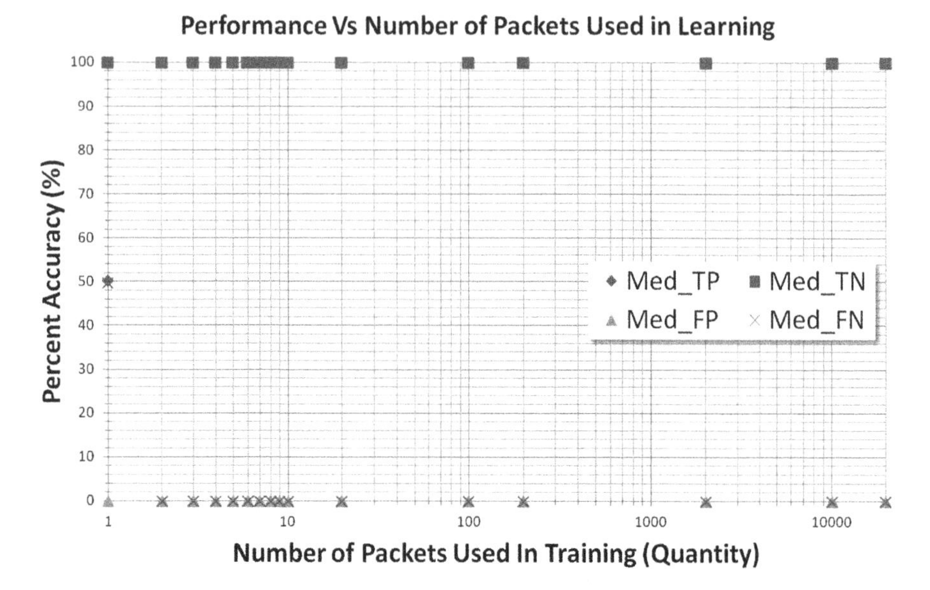

Fig. 2.23 ALIEN Neural network learning classification performance as function of number of number of actual packets used during training

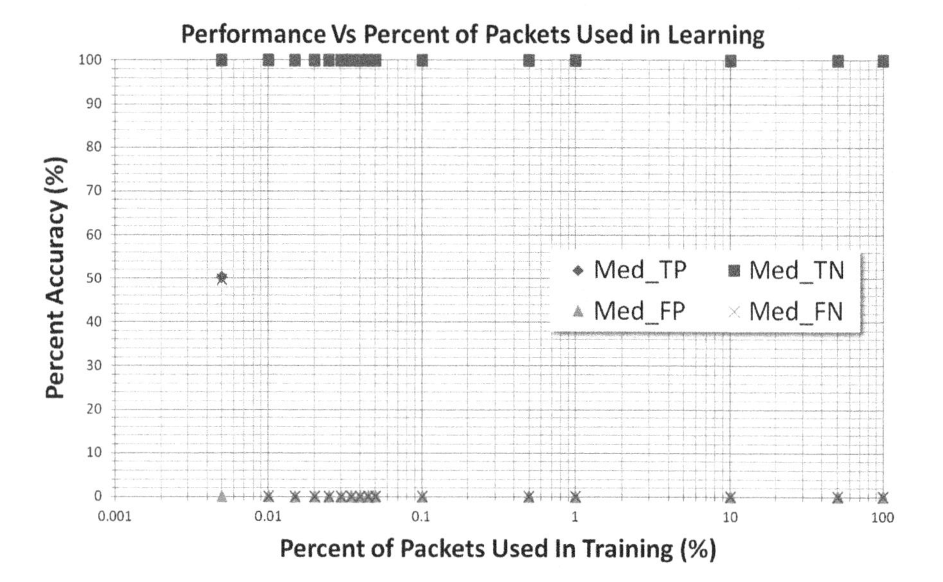

Fig. 2.24 ALIEN Neural network learning classification performance as function of number of number of percent data used during training

Figure 2.23 shows the percent accuracy versus the number of packets used during training. From this Figure, we can observe that if only two packets are used for ALIEN to train the neural network, the classification performance is 100 %. It is interesting to note that if only one packet is used, the true-negative, TN, performance is 100 % while the true-positive, TP, performance is about 50 %. What this specific result highlights is the fact that only one packet of the MX or A type was used during the training. Therefore, the system learned to identify DNS A packet types and guessed with 50 % accuracy on the DNS MX packet type. Figure 2.24 shows the same results of classification accuracy versus the percentage of data used to perform the training of the neural network using ALIEN. The results and figures clearly highlight that given the lowest possible amount of data, 2 packets, the network is able to produce perfect classification results with 100 % accuracy for a total of 20,000 distinct network packets perfectly classified for domain names varying from 5 to 30 bytes in length.

During the previous experiments, the training dataset was randomly chosen. Therefore, it is possible that in some cases, the number of A versus MX packets used during training was greater than the other or vice versa. This would mean the training dataset was not balanced in terms of the representation of the possible lookup types. Thus, we performed an additional experiment in which we ensured that the random dataset selection process included an equal number of A and MX DNS packets. Figure 2.25 shows the experimental results for classifying A and MX packets as function of number of packets used during training. During the experiment, the number of MX and A packets was balanced but the selection of each was random. We performed 100 epochs for each number of packets, and the results reinforce our previous conclusion that ALIEN can achieve 100 % perfect accuracy of classification amongst the 20,000 network packets used during our experiments.

Figures 2.26 and 2.27 display the number of iterations required to achieve perfect classification as function of percent data used and number of packets used. From the

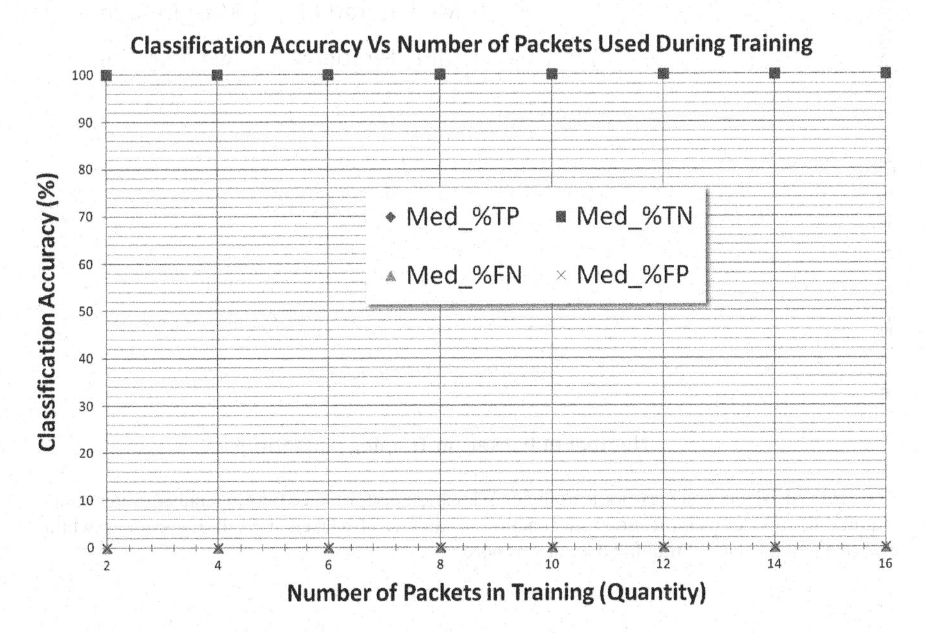

Fig. 2.25 ALIEN Neural network learning classification performance as function of number of number of packets used during training while ensuring that a balanced training dataset was used

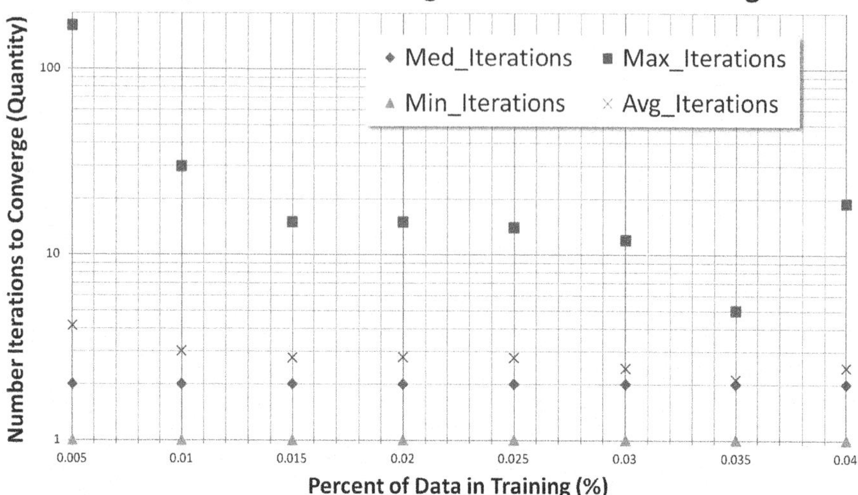

Fig. 2.26 ALIEN neural network learning classification performance for the number of iterations required for perfect classification as function of percent data used during training while ensuring that a balanced training dataset was used

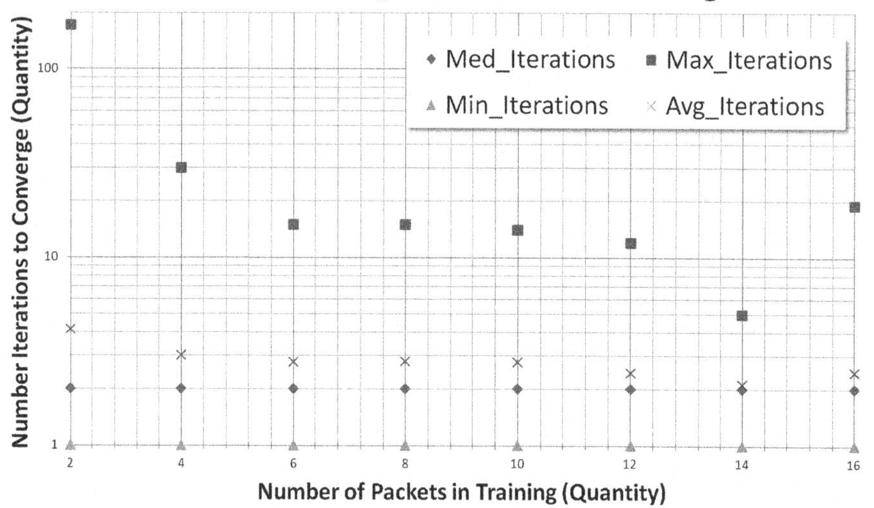

Fig. 2.27 ALIEN neural network learning classification performance for the number of iterations required for perfect classification as a function of number of packets used during training while ensuring that a balanced training dataset was used

results, we can observe that the minimum number of iterations achieved was one single iteration and the maximum covered 100 iterations. In general, the median number of iterations required remained two iterations for all cases during our experiments.

2.7 Conclusions

In this work, we presented a newly developed neural network learning algorithm called ALIEN. The goal of the work is to develop a low architectural overhead learning algorithm to train a memristor-based physical neural network for applications in network security in the tactical environment. This work demonstrated ALIEN's capacity to perform nonlinear classification in the performance of pattern classification. For instance, our experiments yielded 98.46 % detection rate for the KDD99 data set. And in the classification of A and MX DNS record requests, our experiments yielded near perfect results except where the critical information required for classification was missing within the truncated record. This result highlighted a weakness in the ability for a fixed input neural network to be able to effectively classify variable domain name record requests. To solve this challenge, we brokered the problem by dividing DNS QNAME, QTYPE, and QCLASS into three distinct inputs that were easily classified by the neural network. It is evident that this last step trivialized the problem to a three input simple challenge. During this particular version of the experiment, we were able to demonstrate 100 % accuracy classifying 20,000 DNS A and MX records when only two packets were used during training (using one of each A and MX record type).

2.8 Future Work

A potential next step could include porting ALIEN to an FPGA computing architecture were we can boost the computing performance and be able to process an increased number of inputs. We would like to be able to process over 1,000 distinct inputs that would allow for broader classification potential. This will offer the opportunity to test and validate neural network-based network security solutions within an embedded sensor architecture. In addition, we would like to enhance the capabilities of the ALIEN platform to include the ability for the neuron outputs to be analog. This will give us the ability to measure level of confidence in the classification by each neuron and mimic closely the operation of memristor devices within the neural network and characterize the capabilities and limitations of the technology. The idea is for ALIEN to reflect the analog output performance of memristor-based physical neural networks. Also, we would like to evaluate ALIEN's applicability to more practical scenarios such as hierarchical network traffic analysis. The problems may range in granularity from the very specific (such as segregating different DNS request types from amongst each other beyond A and MX DNS records as shown here) to the very general (such as classifying an arbitrary network traffic sample as normal or anomalous based upon an established baseline, similar to the KDD99 challenge). Our hope is that this work will pave the way for the eventual implementation of powerful neuromorphic intrusion detection/prevention solutions in software within a high performance computing cluster formed by multi-core processors, GPUs, or FPGAs and in the future within embedded memristive-based hardware platforms that are small, light, and low in power consumption.

References

1. John Gantz and David Reinsel, "THE DIGITAL UNIVERSE IN 2020: Big Data, Bigger Digital Shadows, and Biggest Growth in the Far East", IDC IVIEW Report, http://www.emc.com/collateral/analyst-reports/idc-the-digital-universe-in-2020.pdf, Sponsored by EMC. Dec. 2012.
2. L.O. Chua, "Memristor - the missing circuit element," IEEE Trans. Circuit Theory, 18 (1971) 507–519.
3. Dmitri B. Strukov, Gregory S. Snider, Duncan R. Stewart, and R. Stanley Williams, "The missing memristor found," Nature 453 (2008), 80–83.
4. R. Williams, "How We Found The Missing Memristor," IEEE Spectrum, 45 (2008) 28-35. http://ieeexplore.ieee.org/stamp/stamp.jsp?tp=&arnumber=4687366&isnumber=4467055
5. R.E Pino, J.W. Bohl, N. McDonald, B. Wysocki, P. Rozwood, K. Campbell, A Timilsina, "Compact method for modeling and simulation of memristor devices: Ion conductor chalcogenide-based memristor devices," 2010 IEEE/ACM International Symposium on Nanoscale Architectures (NANOARCH), Anaheim, CA, June 17-18 (2010) pp 1-4. http://ieeexplore.ieee.org/stamp/stamp.jsp?tp=&arnumber=5510936&isnumber=5510921
6. Chris Yakopcic, Tarek M. Taha, Guru Subramanyam, and Robinson E. Pino, "Memristor SPICE Model and Crossbar Simulation Based on Devices with Nanosecond Switching Time," Proceedings of International Joint Conference on Neural Networks, Dallas, Texas, USA, August 4-9, 2013, pp.
7. C. Yakopcic, T.M. Taha, G. Subramanyam, R.E. Pino, S. Rogers, "A Memristor Device Model," IEEE Electron Device Letters, 32 (2011) 1436-1438.
8. Robert Kozma, Robinson E. Pino, and Giovanni E. Pazienza, "Are Memristors the Future of AI?," R. Kozma et al. (eds.), Advances in Neuromorphic Memristor Science and Applications, Springer, New York, 2012, pp 9-14.
9. Qiangfei Xia, Warren Robinett, Michael W. Cumbie, Neel Banerjee, Thomas J. Cardinali, J. Joshua Yang, Wei Wu, Xuema Li, William M. Tong, Dmitri B. Strukov, Gregory S. Snider, Gilberto Medeiros-Ribeiro, and R. Stanley Williams, "Memristor–CMOS Hybrid Integrated Circuits for Reconfigurable Logic," Nano Letters 9 (2009) 3640-3645.
10. M. Shevenell, "Task 2 - Research Emerging Technologies for Embedded High Performance Intelligent Sensor," ARL internal presentation, Security for Tactical Operations Relying on Methods for Enhancing Robustness (STORMER), 1st Quarter Review 2013 STORMER Team,
11. Joshua S. White, Thomas Fitsimmons, Jeanna N. Matthews, "Quantitative Analysis of Intrusion Detection Systems Snort and Suricata," SPIE Defence and Security, Sensing, Cyber Sensing 2013, Proc. of SPIE 8757 (2013) pp. 875704-1-12.
12. PowerEdge R717 Techical Guide, http://www.dell.com/downloads/global/products/pedge/en/Poweredge-r715-technicalguide.pdf (Viewed on September 12, 2013).
13. FAQs, Raspberry Pi, <http://www.raspberrypi.org/faqs>, (9 September 2013)

14. R. Pino, "Reconfigurable Electronic Circuit," United States Patent 7902857, March 8 (2011).
15. R. Pino, "Neuromorphic Computer," United States Patent 8275728, September 25 (2012).
16. R. Pino, J. Bohl, "Self-Reconfigurable Memristor-Based Analog Resonant Computer," United States Patent 8274312, September 25 (2012).
17. Robinson E. Pino, Youngok K. Pino, "Reconfigurable memristor-based computing logic," United States Patent US 8427203 B2, April 23 (2013).
18. G.S. Rose, R. Pino, Q. Wu, "A low-power memristive neuromorphic circuit utilizing a global/local training mechanism," Proceedings of International Joint Conference on Neural Networks (IJCNN), San Jose, CA, July 31-5, 2011, pp. 2080 – 2086.
19. David Bol, Dina Kamel, Denis Flandre, Jean-Didier Legat, "Nanometer MOSFET effects on the minimum-energy point of 45nm subthreshold logic," Proceedings of the 14th ACM/IEEE international symposium on Low power electronics and design (ISLPED), San Fancisco, CA, August 19–21, 2009, pp. 3-8.
20. R. Pino, H. Li, Y. Chen, M. Hu. B. Liu, "Statistical memristor modeling and case study in neuromorphic computing," 2012 49th ACM/EDAC/IEEE Design Automation Conference (DAC), San Francisco, CA, June 3-7, 585 - 590 (2012).
21. McCulloch and Pitts, "A logical calculus of the ideas immanent in nervous activity," Bulletin of Mathematical Biophysics, 5 (1943) 115-133.
22. O. L. Mangasarian, "Linear and Nonlinear Separation of Patterns by Linear Programming," *Operations Research*, 13 (1965) 444-452.
23. Paul J. Werbos, "Beyond Regression: New Tools for Prediction and Analysis in the Behavioral Sciences," PhD thesis, Harvard University (1974).
24. Paul J. Werbos, "Backpropagation through time: what it does and how to do it," Proceedings of the IEEE, 78 (1990) 1550 – 1560.
25. S. Pervez, I. Ahmad, A. Akram, S. U. Swati, "A Comparative Analysis of Artificial Neural Network Technologies in Intrusion Detections Systems," Proc. of the 6th WSEAS International Conference on Multimedia, Internet & Video Technologies, Lisbon, Portugal, September 22-24 (2006).
26. J.P. Anderson, "Computer Security Threat Monitoring and Surveillance," Technical Report, J.P. Anderson Company, Fort Washington, Pennsylvania April (1980).
27. D. Denning, "An Intrusion-Detection Model," IEEE Tran Software Engineering, SE-13(2) 222-232 (1987).
28. J. Cannady, "Artificial Neural Networks for Misuse Detection," Proc. 21st National Information Systems Security Conference, Arlington, VA, October 5-8 (1998).
29. T. Vollmer, M. Manic, "Computationally Efficient Neural Network Intrusion Security Awareness," Proc. 2nd International Symposium on Resilient Control Systems, Idaho Falls, ID, August 11 - 13 (2009).
30. H.G. Kayacik, A.N. Zincir-Heywood, M.I. Heywood, "On the capability of an SOM based intrusion detection system," 2003 IEEE Proceedings of the International Joint Conference on Neural Networks (IJCNN 2003) July 20-24, 2003, pp.1808,1813
31. KDD Cup 1999 Data, The UCI KDD Archive Information and Computer Science University of California, <http://kdd.ics.uci.edu/databases/kddcup99/kddcup99.html> (9 September 2013).
32. Domain Names - Implementation And Specification, http://www.ietf.org/rfc/rfc1035.txt, (viewed on September 11, 2013)
33. tcpdump, http://www.tcpdump.org/ (viewed on July 23, 2013).

Chapter 3
Automated Cyber Situation Awareness Tools and Models for Improving Analyst Performance

Massimiliano Albanese, Hasan Cam, and Sushil Jajodia

3.1 Introduction

An ever increasing number of critical missions rely today on complex Information Technology infrastructures, making such missions vulnerable to a wide range of potentially devastating cyber-attacks. Attackers can exploit network configurations and vulnerabilities to incrementally penetrate a network and compromise critical systems, thus rendering security monitoring and intrusion detection much more challenging. It is also evident from the ever growing number of high-profile cyber-attacks reported in the news that not only are cyber-attacks growing in sophistication but also in numbers. For these reasons, cyber-security analysts need to continuously monitor large amounts of alerts and data from a multitude of sensors in order to detect attacks in a timely manner and mitigate their impact. However—given the inherent complexity of the problem—manual analysis is labor-intensive and error-prone, and distracts the analyst from getting the "big picture" of the cyber situation.

This chapter explores how automated Cyber Situation (or Situational) Awareness (CSA) tools and models can enhance performance, cognition and understanding for cyber professionals monitoring complex information technology systems. In most current solutions, human analysts are heavily involved in every phase of the

---

The work of Sushil Jajodia and Massimiliano Albanese was supported in part by the Army Research Office under awards W911NF-13-1-0421, W911NF-09-1-0525, and W911NF-13-1-0317, and by the Office of Naval Research under awards N00014-12-1-0461 and N00014-13-1-0703.

M. Albanese (✉) • S. Jajodia
Center for Secure Information Systems, George Mason University, Fairfax, VA 22030, USA
e-mail: malbanes@gmu.edu; jajodia@gmu.edu

H. Cam
Network Science Division, U.S. Army Research Laboratory, Adelphi, MD 20783, USA
e-mail: hasan.cam.civ@mail.mil

R.E. Pino et al. (eds.), *Cybersecurity Systems for Human Cognition Augmentation*, Advances in Information Security 61,
DOI 10.1007/978-3-319-10374-7_3

monitoring and response process. Ideally, we should move from a human-in-the loop scenario to a human-on-the loop scenario, where human analysts have the responsibility to oversee the automated processes and validate the results of automated analysis of monitoring data. This may greatly enhance not only their performance—which could be measured, for instance, in terms of false alarms and missed detections—but also their cognition and understanding of the cyber situation. In fact, in the human-on-the loop scenario, analysts would be assigned higher-level analysis tasks, whereas in most current scenarios they are responsible for analyzing large volumes of fine-grained data. In order to assess the performance of task assignment of analysts in the human-on-the loop scenario, it is highly desirable to have a temporal model such as Petri net to represent and integrate the concurrent operations of cyber-physical systems with the cognitive processing of analyst.

The chapter is organized as follows. Section 3.2 introduces the fundamental definition of situation awareness in the context of cyber defense. Section 3.3 identifies some of the key questions that would greatly benefit from the adoption of automated CSA tools, whereas Sect. 3.4 provides an overview of the state of the art. Then, Sect. 3.5 describes in more detail how automated Cyber Situation Awareness (CSA) tools can be integrated into a coherent framework and how they can enhance performance, cognition and understanding for cyber professionals. Section 3.6 presents new Petri net models to identify and integrate the concurrent operations of cyber sensors together with analyst cognitive processes within a cyber situation awareness model. Finally, concluding remarks are given in "Conclusions".

3.2 Definition of Cyber Situational Awareness

A multidisciplinary group of leading researchers from cyber security, cognitive science, and decision science have elaborated on the fundamental challenges facing the research community and have identified promising solution paths [4].

Today, when a security incident occurs, the top three questions security administrators would ask are in essence: What happened? Why did it happen? What should I do? Answers to the first two questions form the core of Cyber Situational Awareness. Whether the last question can be satisfactorily answered is greatly dependent upon the cyber situational awareness capabilities of an enterprise.

A variety of computer and network security research topics belong to or touch the scope of Cyber Situational Awareness. However, the Cyber Situational Awareness capabilities of an enterprise are still very limited for several reasons, including, but not limited to:

- Inaccurate and incomplete vulnerability analysis, intrusion detection, and forensics.
- Lack of capability to monitor certain microscopic system or attack behaviors.
- Limited capability to transform/fuse/distill information into cyber intelligence.
- Limited capability to handle uncertainty.
- Existing system designs are not very "friendly" to Cyber Situational Awareness.

Without losing generality, cyber situation awareness can be viewed as a three-phase process: situation perception, situation comprehension, and situation projection. *Perception* provides information about the status, attributes, and dynamics of relevant elements within the environment. *Comprehension* of the situation encompasses how people combine, interpret, store, and retain information. Projection of the elements of the environment (situation) into the near future encompasses the ability to make predictions based on the knowledge acquired through perception and comprehension. Situation awareness is gained by a system, which is usually the system being threatened by random or organized cyber-attacks.

Several definitions of Cyber Situational Awareness have been proposed in the literature. In [7], Endsley provides a general definition of Situation Awareness (SA) in dynamic environments:

> *Situation awareness is the perception of the elements of the environment within a volume of time and space, the comprehension of their meaning, and the projection of their status in the near future.*

Endsley also differentiates between situation awareness—which he defines as "*a state of knowledge*"—and situation assessment—which he defines as a "*process of achieving, acquiring, or maintaining situation awareness.*" This distinction becomes extremely important when trying to apply computer automation to situation awareness. Since situation awareness is "a state of knowledge", it is a cognitive process and resides primarily in the minds of human analysts, while situation assessment—as a process or set of processes—lends itself to automated techniques. Another definition of situational awareness is provided by Alberts et al. in [8]:

> *When the term situational awareness is used, it describes the awareness of a situation that exists in part or all of the battlespace at a particular point in time. In some instances, information on the trajectory of events that preceded the current situation may be of interest, as well as insight into how the situation is likely to unfold. The components of a situation include missions and constraints on missions (e.g., Rules of Engagement), capabilities and intentions of relevant forces, and key attributes of the environment.*

The Army Field Manual 1-02 (September 2004) defines Situational Awareness as:

> *Knowledge and understanding of the current situation which promotes timely, relevant and accurate assessment of friendly, competitive and other operations within the battlespace in order to facilitate decision making. An informational perspective and skill that fosters an ability to determine quickly the context and relevance of events that are unfolding.*

There are many other definitions available for situation awareness but the ones described above seem to be or are becoming widely accepted.

The goal of current research in this area is to explore ways to elevate the Cyber Situational Awareness capabilities of an enterprise to the next level by measures such as developing holistic Cyber Situational Awareness approaches and evolving existing system designs into new systems that can achieve self-awareness. One major output of current efforts is a set of scientific research objectives and challenges in the area of Cyber Situational Awareness, which can be represented in terms of a set of key questions to be addressed, as discussed in more detail in the following section.

3.3 Some Key Questions

In order to effectively monitor complex cyber systems, analysts need to analyze large volumes of data and formulate hypotheses, validate such hypotheses, and ultimately find answers to a number of highly critical questions that will guide any subsequent decision process, including intrusion response and mitigation. Such questions cover a broad range of aspects, including current situation, impact and evolution of an attack, behavior of the attackers, forensics, quality of available information and models, and prediction of future attacks. Below are some of the key questions that would greatly benefit from the adoption of automated CSA tools.

Current situation and evolution

- *Is there any ongoing attack?*
- *If yes, where is the attacker?*
- *How is the situation evolving?*

Answering this set of questions implies the capability of effectively detecting and monitoring ongoing intrusions, and identifying the assets that may have been compromised. This aspect can also be called *situation perception*. Situation perception includes both *situation recognition* and *identification*. Situation identification can include identifying the type of attack, the source of an attack, and the target, whereas recognition is only acknowledging that an attack is occurring. Situation perception is beyond intrusion detection. An Intrusion Detection System (IDS) simply identifies events that may be part of an attack.

This is certainly one of the key areas where analysts are most likely to struggle in dealing with large volumes of data and trying to correlate individual events and alerts into multi-step attack scenarios. Therefore, automated tools would improve performance by providing analysts with a set of hypotheses, which they can simply validate based on their knowledge and experience.

Impact

- *How is the attack impacting the enterprise or mission?*
- *How can we assess the damage?*

Answering this set of questions implies the capability of accurately assessing the impact of ongoing attacks. There are two parts to *impact assessment*: (i) assessment of current impact (damage assessment); and (ii) assessment of future impact (additional damage that would occur if the attacker continues on this path or, more generally, if the activity of interest continues). Vulnerability analysis is also largely an aspect of impact assessment, as it provides knowledge of "us"—the assets being defended—and enables projection of future impact. Assessment of future impact also involves threat analysis.

Analyst performance and cognition could be enhanced by formal metrics to assess impact and damage. This would relieve analysts from time-consuming assessments and provide them with a better understating of the current situation.

Attacker's behavior and prediction

- *How are the attackers expected to behave?*
- *What are their strategies?*
- *Can we predict plausible futures of the current situation?*

Answering this set of questions implies the capability of modeling the attacker's behavior, in order to understand its goals and strategies, and predicting possible future moves. A major component of this aspect is attack trend and intent analysis, which are more oriented towards the behaviors of an adversary or actor(s) within a situation than with the situation itself.

Understanding an attacker's behavior is a complex task that requires integrating knowledge of the target system, knowledge of the attacker, as well as many other sources of information. Additionally, this task requires deductive skills that are difficult to automate. Nonetheless, the analyst's cognitive processes could benefit from tools capable of defining formal models of the attackers and formulate hypotheses accordingly.

Forensics

- *How did the attacker create the current situation?*
- *What was he trying to achieve?*

Answering this set of questions implies the capability of analyzing available logs and correlating observations in order to understand how attacks evolved. This aspect includes causality analysis. Clearly, this is another area where, given the large volumes of data involved, analysts could greatly benefit from automated tools.

3.4 State of the Art

Although the ultimate goal of research in Cyber Situation Awareness is to design systems capable of gaining self-awareness—and do self-protection—without involving any humans in the loop, this vision is still very distant from the current reality, and there still does not exist a tangible roadmap to achieve this vision in a practical way.

In this chapter, we view human analysts and decision makers as an indispensable *component* of the system gaining situation awareness. Nonetheless, we show that humans in the loop can greatly benefit from the adoption of automated tools capable of reducing the semantic gap between an analyst's cognitive processes and the huge volume of available fine-grained monitoring data.

Practical cyber situational awareness systems include not only hardware sensors (e.g., a network interface card) and "smart" computer programs (e.g., programs that can learn attack signatures), but also mental processes of human beings making advanced decisions [5, 6]. Cyber situation awareness can be gained at multiple abstraction levels: raw data is typically collected at the lower levels, whereas more refined information is collected at the higher levels, as data is analyzed and converted into more abstract information. Data collected at the lowest levels can easily overwhelm the cognitive capacity of human decision makers, and situation awareness based solely on low level data is clearly insufficient.

Cyber SA systems and Physical SA systems have fundamental differences. For instance, Physical SA systems rely on specific hardware sensors and sensor signal processing techniques, but neither the physical sensors nor the specific signal processing techniques play an essential role in Cyber SA systems (although there is research that has looked at applying signal processing techniques to analyze network traffic and trends). Cyber SA systems rely on cyber sensors such as IDS', log files, anti-virus systems, malware detectors, and firewalls; they all produce events at a higher level of abstraction than raw network packets. Additionally, the speed at which the cyber situation evolves is usually orders of magnitude higher than in physical situation evolution.

Existing approaches to gain cyber situation awareness consist of vulnerability analysis (using attack graphs), intrusion detection and alert correlation, attack trend analysis, causality analysis and forensics (e.g., backtracking intrusions), taint and information flow analysis, damage assessment (using dependency graphs), and intrusion response. These approaches however only work at the lower (abstraction) levels. Higher level situation-awareness analyses are still done manually by a human analyst, which makes it labor-intensive, time-consuming, and error-prone.

Although researchers have recently started to address the cognitive needs of decision makers, there is still a big gap between human analysts' mental models and the capability of existing cyber situation-awareness tools.

Existing approaches need to handle uncertainty better. Uncertainty in perceived data could lead to distorted situation awareness. For example, attack graph analysis toolkits are designed to do deterministic attack consequence estimation. In real time cyber situation awareness, such consequence estimates could be very misleading due to various uncertainties. Alert correlation techniques cannot handle the inherent uncertainties associated with inaccurate interpretations of intrusion detection sensor reports (such inaccurate interpretations lead to false positives/negatives in determining whether an IDS alert corresponds to an attack).

Lack of data and incomplete knowledge may raise additional uncertainty management issues. For example, lack of data leads to insufficient understanding of the system being defended. Such incompleteness may be caused by imperfect information about system configurations, incomplete sensor deployment, etc.

Existing approaches also lack the reasoning and learning capabilities required to gain full situation awareness for cyber defense. The key questions of cyber situation awareness (see Sect. 3.3) have been treated as separate problems, but full cyber situation awareness requires all these aspects to be integrated into one solution. Such a solution is in general still missing, but the framework discussed in Sect. 3.5 represents a first important step in this direction. Furthermore, looking beyond cyber SA and considering how cyber SA solutions complement the other cyber defense technologies, cyber SA activities need to be better integrated with effect-achieving or environment-influencing activities (e.g., intrusion response activities).

3.5 Automated Cyber Situation Awareness

In this section, we present a number of techniques and automated tools that can be jointly used to enhance an analyst's performance as well as his understanding of the cyber situation. Most of the work presented in this section is the outcome of research efforts conducted by the authors as part of a funded multi-year multi-university research project.

The first step in achieving any level of automation in the situation awareness process is to develop the capability of modeling cyber-attacks and their consequences. Attack graphs have been widely used to model attack patterns, and to correlate alerts [9–13]. However, existing approaches typically have two major limitations. First, attack graphs do not provide mechanisms for evaluating the likelihood of each attack pattern or its impact on the enterprise or mission. Second, scalability of alert correlation has not been fully addressed. In order to address these limitations, we have defined a framework to analyze massive amounts of raw security data in real time, and assess the impact of current and future attacks. The proposed framework ultimately presents the analyst with a number of plausible scenarios, thus relieving him from the daunting task of mining large amounts of raw data.

The proposed framework is illustrated in Fig. 3.1. We start from analyzing the topology of the network, known vulnerabilities, zero-day vulnerabilities, and their interdependencies. Vulnerabilities are often interdependent, making traditional point-wise vulnerability analysis ineffective.

Our topological approach to vulnerability analysis allows to generate accurate attack graphs showing all the possible attack paths within the network. A node in an attack graph represents (depending on the level of abstraction) a subnet, an individual machine, an individual software application, or an individual vulnerability, each of which can be exploited by an attacker. An example of attack graph is shown in Fig. 3.2. An edge from a node, say A, to another node, say B, represents the fact that B can be exploited after A, and it is labeled with the probability that an occurrence of the attack will exploit B within a given time period after A. This approach extends the classical definition of attack graph by encoding probabilistic knowledge of the attacker's behavior.

In order to enable concurrent monitoring of multiple attack types, we merge multiple attack graphs in a compact data structure and define an index structure on

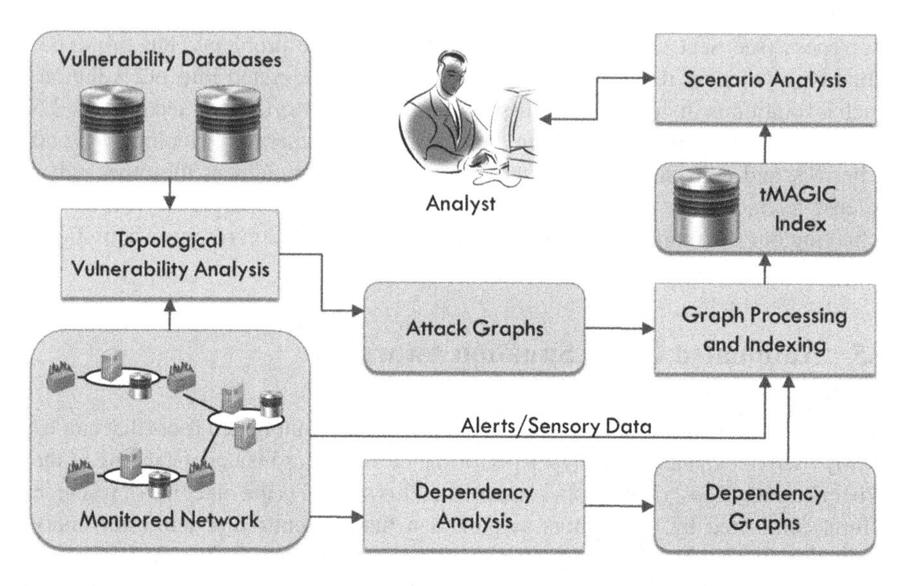

Fig. 3.1 Cyber situation awareness framework

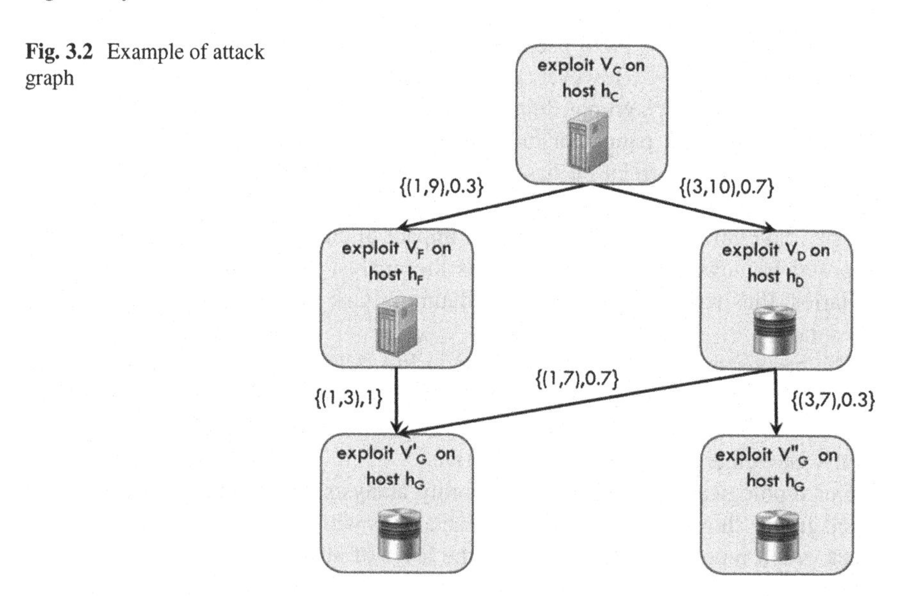

Fig. 3.2 Example of attack graph

top of it to index large amounts of alerts and sensory data (events) in real-time [2, 3]. The proposed index structure allows us to solve three important problems:

- Given a sequence of events, a probability threshold, and an attack graph, find all minimal subsets of the sequence that validate the occurrence of the attack with a probability above the threshold (evidence);
- Given a sequence of events and a set of attack graphs, identify the most likely type of attack occurring in the sequence (identification);
- Identify all possible outcomes of the current situation and their likelihood (prediction).

We also perform dependency analysis to discover dependencies among services and/or machines and derive dependency graphs encoding how these elements depend on one other [2]. Dependency analysis is fundamental for assessing current damage caused by ongoing attacks (value/utility of services disrupted by the attacks) and future impact of ongoing attacks (value/utility of additional services that would be disrupted if no action is taken). In fact, any service or machine within a complex system may be indirectly affected by a successful exploit against another service or machine it relies upon.

For each possible outcome of the current situation, we can then compute a future damage assessment by introducing the notion of attack scenario graph, which combines dependency and attack graphs, thus bridging the gap between known vulnerabilities and the missions or services that could be ultimately affected. An example of attack scenario graph is shown in Fig. 3.3. In the figure, the graph on the left is an attack graph modeling all the vulnerabilities in the system and their relationships, whereas the graph on the right is a dependency graph capturing all the explicit and implicit dependencies between services and machines. The edges from nodes in the attack graph to nodes in the dependency graph indicate which services or machines are directly impacted by a successful vulnerability exploit, and are labeled with the corresponding exposure factor, that is the percentage loss the affected asset would experience upon successful execution of the exploit.

Finally, in [2], we proposed efficient algorithms for both detection and prediction, and showed that they scale well for large graphs and large volumes of alerts.

In summary, the proposed framework provides security analysts with a high-level view of the cyber situation. From the simple example of Fig. 3.3—which models a system including only a few machines and services—it is clear that manual analysis could be extremely time-consuming even for relatively small systems. Instead, the graph of Fig. 3.3 provides analysts with a visual and very clear understanding of the

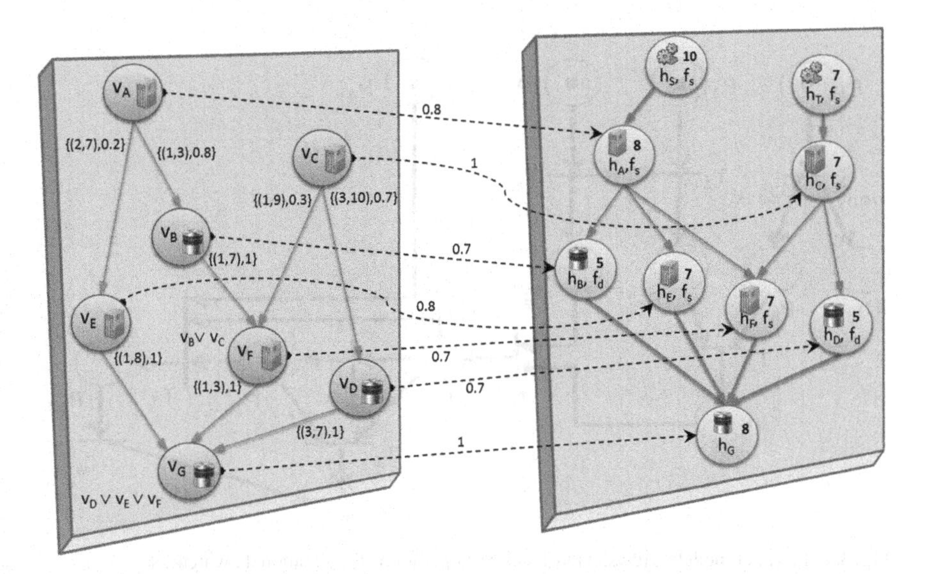

Fig. 3.3 Example of attack scenario graph

situation, thus enabling them to focus on higher-level tasks that require experience and intuition, and thus are more difficult to automate. Additionally, other classes of automated analytical processes may be developed within this framework to support the analyst even during these higher-level tasks. For instance, based on the model of Fig. 3.3, we could automatically generate a ranked list of recommendations on the best course of actions analysts should take to minimize the impact of ongoing and future attacks (e.g., sets of network hardening actions [1]).

3.6 Petri Net Models for Situational Awareness

While dealing with concurrent operations of vulnerability exploitation, impact and recovery of attacks in the human-on-the loop scenario, it is desirable to have an integrated temporal model for characterizing them along with analyst cognitive processes. Petri nets have been proven capable of modeling the concurrent operations of firewall, vulnerability exploitation, and recovery of attacks in temporal domain of cyber-physical systems [14]. This section extends this model to include the cognitive processes of cyber analysts. Moreover, we show how Bayesian networks can be employed to incorporate uncertainties and analyst cognition into the process of determining the actual state of a system.

3.6.1 Petri Net Model of Physical Systems and Operations

This section first introduces a time Petri net model, shown in Fig. 3.4, to integrate the operations of firewall, sensor measurements, vulnerability scanning, and recovery.

The semantics of places and transitions in the Petri Net of Fig. 3.4 is as follows.

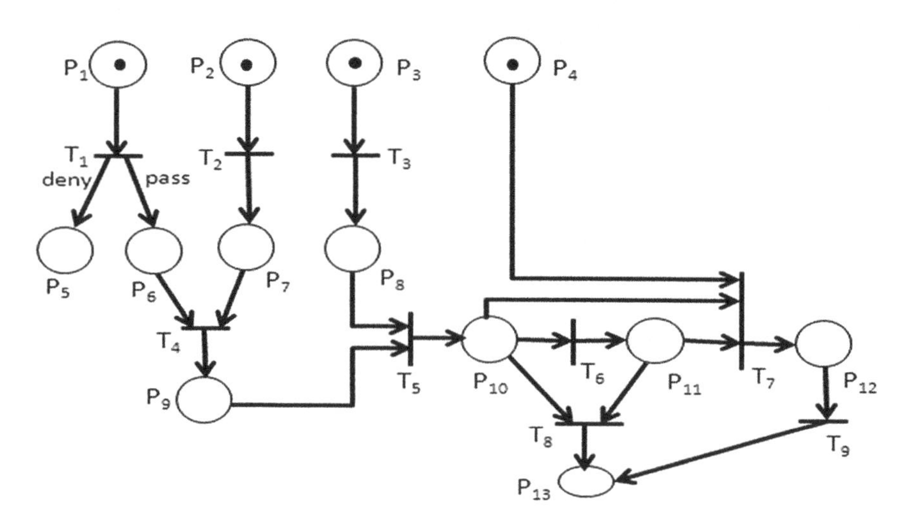

Fig. 3.4 Petri net model of physical systems and operations for situational awareness

- P_1: Firewall receives packets.
- P_2: Sensors' measurements are collected.
- P_3: Vulnerability scanner scans.
- P_4: Recovery tools run.
- P_5: Reject firewall ruleset-matched packets.
- P_6: Pass ruleset-non-matched packets (benign & malicious).
- P_7: Attackability conditions of system.
- P_8: Vulnerabilities exist.
- P_9: Active malicious codes.
- P_{10}: Assets compromised.
- P_{11}: Impact of assets damages.
- P_{12}: Assets recovered partially.
- P_{13}: Available assets.
- T_1: Apply firewall ruleset against packets.
- T_2: Alarm probability exceeds threshold.
- T_3: Find new vulnerabilities.
- T_4: Activated malicious packets.
- T_5: Intrusion attempts.
- T_6: Propagate impact of damages.
- T_7: Patch vulnerabilities, and recover damages.
- T_8: Evict compromised non-recoverable assets.
- T_9: Recover assets fully.

The places of tokens at a given time indicate a particular status of these operations, assisting to recognize the cyber situation of the distributed control system. For instance, the transition T_1 denotes the event of running firewall ruleset against incoming packets and, then, those packets matching firewall rules are denied and the others are allowed to pass. These latter packets may contain malicious data and could be activated in transition $T_{4,}$ depending on the status of system attackability conditions and the value of alarm thresholds in T_2. When malicious data match vulnerabilities in T_5, vulnerabilities are exploited, leading to the damage on assets as illustrated by P_{10}. The propagation of damage is indicated by firing transition T_6. To repair the damages, recovery and/or resilience mechanisms can be activated by triggering transition T_7. The availability status of all assets is represented by P_{13}. Note that tokens dynamically move around the places of the model as its transitions are fired. Therefore, the marking of tokens in the model helps determine the cyber situational status of a distributed control system.

3.6.2 Incorporating Analyst Cognitive Processes into the Petri Net Model

The analyst cognitive processing of observation-hypothesis-action is integrated with the Petri net model of physical devices, as shown in Fig. 3.5. After observing all the information obtained through physical devices, an analyst comprehends the

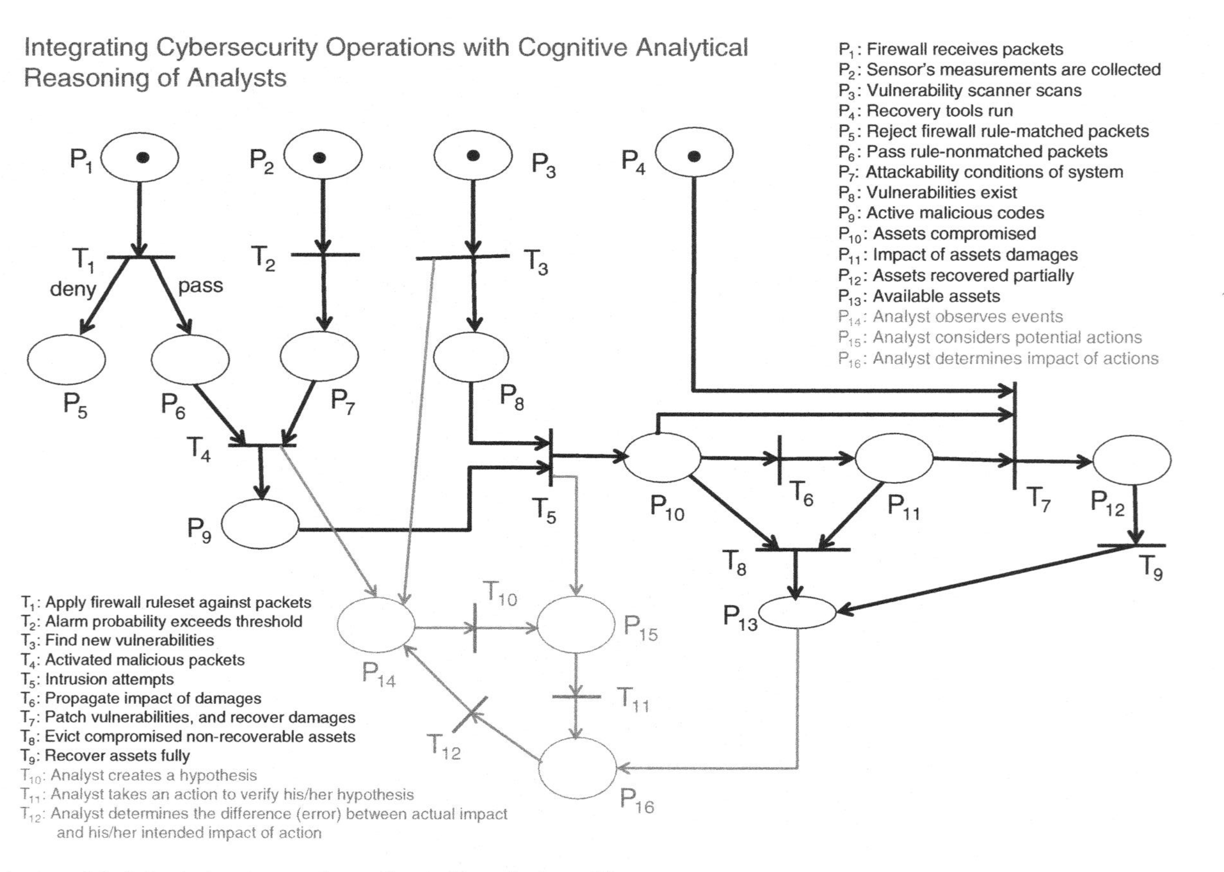

Fig. 3.5 Petri net model of physical systems and operations with analyst cognition process

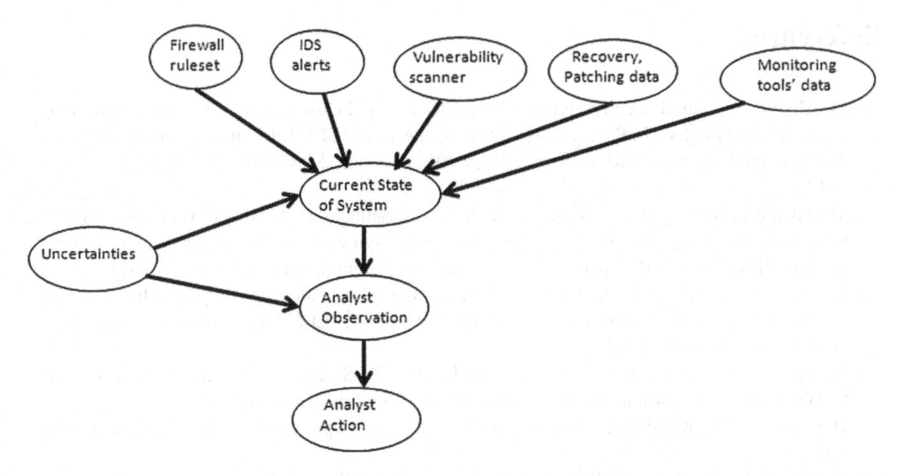

Fig. 3.6 Bayesian network for an integrated cognitive processing and physical systems with uncertainties

situation in his/her capacity and develops a hypothesis for identifying any existing cybersecurity problem in the system, and then takes an action to resolve the problem under the guidance of the hypothesis. If the action fails or is not sufficient to fix the problems, then the analyst observes again the situation along with uncertainty information [15], develops a new hypothesis, and takes a corresponding action, which is repeated over time as many times as needed.

A Bayesian network will be used to determine iteratively the current state of the system by incorporating dynamic impact of uncertainties and analyst observation/action into the information obtained via physical devices, as illustrated in Fig. 3.6.

3.7 Conclusions

In this chapter, we have highlighted the importance of developing automated tools and models to support the work of security analysts for cyber situation awareness. Current processes are mostly manual and ad-hoc, therefore they are extremely time-consuming and error-prone, and force analysts to seek through large amounts of fine-grained monitoring data, rather than focusing on the big picture of the cyber situation. To address this limitation, we have shown how an integrated set of automated tools can be used to perform a number of highly repetitive and otherwise time-consuming tasks in a highly efficient and effective way. The result of this type of automated analysis is the generation of a set of higher-level attack scenarios that can be used by analysts to assess the current situation as well as to project it in the near future. We believe this is an important step towards future generations of self-aware and self-protecting systems, but more work needs to be done in this direction to achieve this vision.

References

1. M. Albanese, S. Jajodia, and S. Noel. "Time-Efficient and Cost-Effective Network Hardening Using Attack Graphs". In Proceedings of the 42nd Annual IEEE/IFIP International Conference on Dependable Systems and Networks (DSN 2012), Boston, Massachusetts, USA, June 25-28, 2012.
2. M. Albanese, S. Jajodia, A. Pugliese, and V. S. Subrahmanian. "Scalable Analysis of Attack Scenarios". In Proceedings of the 16th European Symposium on Research in Computer Security (ESORICS 2011), pages 416-433, Leuven, Belgium, September 12-14, 2011.
3. M. Albanese, A. Pugliese, and V. S. Subrahmanian. "Fast Activity Detection: Indexing for Temporal Stochastic Automaton based Activity Models". IEEE Transactions on Knowledge and Data Engineering, 2013.
4. "Cyber Situational Awareness: Issues and Research". S. Jajodia, P. Liu, V. Swarup, and C. Wang (Eds.), Vol. 46 of Advances in Information Security, Springer, 2010.
5. H. Gardner. "The Mind's New Science: A History of the Cognitive Revolution", Basic Books, 1987.
6. P. Johnson-Laird, "How We Reason", Oxford University Press, 2006.
7. M. Endsley. "Toward a theory of situation awareness in dynamic systems". In Human Factors Journal, volume 37(1), pages 32–64, March 1995.
8. D. S. Alberts, J. J. Garstka, R. E. Hayes, and D. A. Signori. "Understanding information age warfare". In DoD Command and Control Research Program Publication Series, 2001.
9. P. Ammann, D. Wijesekera, and S. Kaushik, "Scalable, graph-based network vulnerability analysis," in Proceedings of the 9th ACM Conference on Computer and Communications Security (CCS 2002), pp. 217–224, Washington, DC, USA, November 2002.
10. C. Phillips and L. P. Swiler, "A graph-based system for network-vulnerability analysis," in Proceedings of the New Security Paradigms Workshop (NSPW 1998), pp. 71–79, Charlottesville, VA, USA, September 1998.
11. S. Jajodia, S. Noel, P. Kalapa, M. Albanese, and J. Williams, "Cauldron: Mission-centric cyber situational awareness with defense in depth," in Proceedings of the Military Communications Conference (MILCOM 2011), Baltimore, MD, USA, November 2011.
12. L. Wang, A. Liu, and S. Jajodia, "Using attack graphs for correlating, hypothesizing, and predicting intrusion alerts," Computer Communications, vol. 29, no. 15, pp. 2917–2933, September 2006.
13. M. Albanese, S. Jajodia, A. Singhal, and L. Wang. "An Efficient Approach to Assessing the Risk of Zero-Day Vulnerabilities". In Proceedings of the 10th International Conference on Security and Cryptography (SECRYPT 2013), Reykjavìk, Iceland, July 29-31, 2013.
14. H. Cam, P. Mouallem, Y. Mo, B. Sinopoli, and B. Nkrumah, "Modeling Impact of Attacks, Recovery, and Attackability Conditions for Situational Awareness", *Proc. of 2014 IEEE International Multi-Disciplinary Conference on Cognitive Methods in Situation Awareness and Decision Support (CogSIMA)*, March 3-6, 2014, San Antonio, TX, USA.
15. P. Xie, J.H. Li, X. Ou, P. Liu, and R. Levy, "Using Bayesian Networks for Cyber Security Analysis," Proc. of 2010 IEEE/IFIP International Conference on Dependable Systems and Networks (DSN), 2010.

Chapter 4
Data Mining in Cyber Operations

Misty Blowers, Stefan Fernandez, Brandon Froberg, Jonathan Williams, George Corbin, and Kevin Nelson

4.1 Introduction

Cyber operations has been roughly defined as the employment of cyber capabilities to achieve military objectives or effects in or through cyberspace [1]. Defending cyberspace is a complex and largely scoped challenge which considers emerging threats to security in space, land, and sea. The *Joint Publication 1-02*, Department of Defense (DoD) *Dictionary of Military and Associated Terms* defines cyberspace as a global domain within the information environment consisting of the interdependent network of information technology infrastructures, including the Internet, telecommunications networks, computer systems, and embedded processors and controllers [1]. Cyberspace operations is defined as the employment of cyber capabilities where the primary purpose is to achieve military objectives or effects in or through cyberspace. Such operations include computer network operations and activities to operate and defend the Global Information Grid. The global cyber infrastructure presents many challenges because of the complexity and massive amounts of information transferred across the global network daily. The cyber infrastructure is a made up of the data resources, network protocols, computing platforms, and computational services that bring people, information, and computational tools together. Techniques from the field of data mining can be used to help make sense of the massive amounts of data that make up the cyber infrastructure.

M. Blowers (✉) • S. Fernandez • B. Froberg • J. Williams
Air Force Research Laboratory, Information Directorate, Rome, NY 13441, USA
e-mail: misty.blowers@us.af.mil

G. Corbin • K. Nelson
BAE Systems, Rome, NY 13440, USA

R.E. Pino et al. (eds.), *Cybersecurity Systems for Human Cognition Augmentation*, Advances in Information Security 61,
DOI 10.1007/978-3-319-10374-7_4

4.2 Data Mining

According to Han and Kamber [2], data mining is a process of discovering interesting patterns in large amounts of data which as previously noted is often a challenge in cyber operations. In order to gain a tactical edge, a warfighter must be able to apply data mining techniques to be maneuverable in cyber space. Maneuverability in cyberspace allows attackers and defenders to simultaneously conduct actions across multiple systems at multiple levels of warfare. For defenders, this can mean hardening multiple systems simultaneously when new threats are discovered, killing multiple access points during attacks, collecting and correlating data from multiple sensors in parallel or other defensive actions [3]. The complexity and dynamics of cyber operations is only weakly understood, especially when a nation is engaged in cyber-warfare.

The dynamic nature of the cyberspace environment presents opportunities for both attackers and defenders to conduct complex cyber operations in serial or parallel across multiple networks and systems [4]. Defensive operators must be vigilant to identify new attack vectors, real-time attacks as they happen, and signs of attacks that have gotten through the security perimeter. This means that defenders must continuously sift through vast amounts of sensor data that could be made more efficient with advances in data mining techniques to accurately map the attack surface, collect and integrate data, synchronize time, select features, develop models, extract knowledge and produce useful visualization. Effective techniques would enable models that describe dynamic behavior of complicated attacks and failures and allow defenders to detect and differentiate simultaneous sophisticated attacks on a target network [4]. Defensive operators that manage an enterprise-level network, distributed networks or multiple, interoperating networks face a significant challenge of strategic coordination to defend against complex cyber-attacks. These operators clearly face a "big data" problem [5].

"Big Data" is about the growing challenge in how we deal with the large and fast-growing sources of data or information. It presents a complex range of analysis and use problems [6]. There are many considerations when dealing with massive amounts of data. One challenge is in having a computing infrastructure that can ingest, validate, and analyze high volumes (size and/or rate) of data. Another is in assessing mixed data (structured and unstructured) from multiple sources. It is often very challenging to deal with unpredictable content with no apparent schema or structure, and often a challenge enabling real-time or near-real-time collection, analysis, and answers [6].

Before one attempts to extract useful knowledge from data, it is important to understand the steps in the data mining process. Simply knowing many algorithms used for data analysis is not sufficient for successful data mining (DM). The figure below outlines the process of mining data that leads to knowledge discovery.

Fayyad et al. [35] describe the knowledge discovery from data model as a series of nine steps Fig 4.1.

1. Develop and understand the application domain. This step includes learning the relevant prior knowledge and considers the goals of the end user.

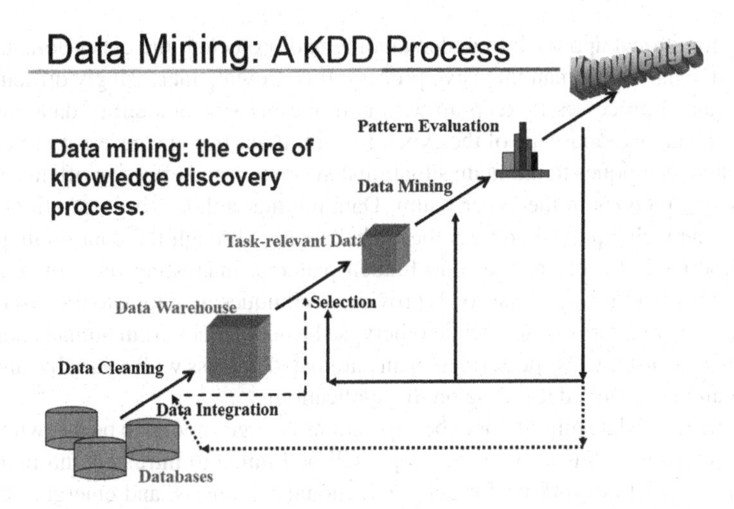

Fig. 4.1 The Knowledge Discovery from Data process allows for the "mining" of valuable knowledge from vast amounts of data just as a miner mines for gold [2]

2. Create a target data set. Here the data miner selects a subset of variables (attributes or features) and data points (examples) that will be used to perform discovery tasks. This step usually includes querying the existing data to select the desired subset.
3. Data cleaning and preprocessing. This step consists of considering outliers, dealing with noise and missing values in the data, and accounting for time sequence information and known changes. Outliers may be irrelevant or be significantly relevant depending on the task at hand.
4. Data reduction and projection. This step consists of finding useful attributes by applying dimension reduction and transformation methods, and finding invariant representation of the data.
5. Choosing the data mining task. Here the data miner matches the goals defined in Step 1 with a particular DM method, such as classification, regression, clustering, etc.
6. Choosing the data mining algorithm. The data miner selects methods to search for patterns in the data and decides which models and parameters of the methods used may be appropriate.
7. Data mining. This step generates patterns in a particular representational form, such as classification rules, decision trees, regression models, trends, etc. More advanced machine learning methods also may apply here.
8. Interpreting mined patterns. Here the analyst performs visualization of the extracted patterns and models, and visualization of the data based on the extracted models.
9. Consolidating discovered knowledge. The final step consists of incorporating the discovered knowledge into the performance system, and documenting and reporting it to the interested parties. This step may also include checking and resolving potential conflicts with previously believed knowledge. In the cyber domain, metrics to measure the effectiveness of detection or battle damage assessment is considered.

The traditional approach to understanding and protecting the cyber domain is a highly manual and human intensive process. It is growing increasingly difficult for these manual processes to keep up with both the massive amount of data and the quickly changing landscape of the cyber domain. It has become necessary to utilize automated techniques to maintain situational awareness and effective offensive and defensive strategies in the cyber realm. Data mining within cyber operations provides some techniques to address these challenges. Through the data mining process described above, one can find hidden patterns, interesting data, or relevant correlations within large datasets. It provides techniques to automate the discovery of structure or patterns which would otherwise be out of reach from human analysts. This analysis is typically performed in an automated process with a variable amount of human interaction, depending on the application.

The scope of data mining for cyber operations is large enough to be its own book, so for purposes of this chapter the scope will be limited to intrusion and malware detection, social networking for cyber situational awareness, and emerging topics for data mining in cyber operations.

4.3 Data Mining for Intrusion Detection

Intrusion Detection and Prevention Systems (IDPS) are automated software designed to monitor traffic or mine through select data sources in search of evidence of an intruder attempting to compromise the network. An IDPS is created to monitor characteristics of a host, the network, and combination of both host/network [9]. IDPSs use three basic types of detection to discover intrusions: signature-based detection, anomaly-based detection, and stateful protocol analysis [10].

Signature-based IDPS use signatures, patterns known to indicate a threat, to compare to observable event patterns in order to identify a current threat [10]. A signature-based IDPS is used in firewalls as a first line of defense as it can efficiently identify threats and act before damage is done for very precisely defined and common threats. A disadvantage to this approach is that it relies entirely on a database of known attack signatures to compare against the current network activity. Data mining may be applied to a signature-based IDPS by observing and analyzing known and suspected attacks to discover new signatures and patterns indicative of an intrusion [11].

Applying data mining techniques allows not only for these previously undiscovered signatures to be found, but also for generalized patterns of attacks to be seen. New and novel attacks, which may not exactly match a previously observed signature, may still match the general patterns of an attack that were learned through data mining techniques.

Anomaly-based detection depends on understanding normal patterns of network activity and looking for activity which appears abnormal relative to normal activity [10]. The vast majority of new threats will come in as anomalous traffic and yet will likely be undetectable by Signature-based Detectors until new signature rules can be created once they are detected, countered and accounted for in the signature database. An anomaly-based IDPS can be successful in detecting attacks which are novel or vary too far from a signature to be detectable by the signature-based IDPS. They are slow to train and heavily dependent upon having very good "normal" data to upon which to base the training. Data mining is very applicable to this approach, as anomaly detection relies entirely on defining a baseline of normalcy. Various data mining techniques may be effectively used to learn a meaningful definition of normalcy based on known benign network connections [12].

Stateful Protocol Analysis also looks at behavior outside of known signature patterns to precisely how protocols are designed to be used and what the protocol creators expect to see when those protocols are used [10]. The key is not only in finding anomalous behavior, but also in finding an anomalous behavior beyond what is typical for a specific network activity. Part of understanding a stateful interaction between a user and a network resource is the series of communications between them and not just individual packets as signature-based and most anomaly-based detectors are usually looking at. Looking at the state of the transactions, the intent of the user is revealed. Monitoring state in a network is complex and requires a lot of processing power in high volume networks. As new normal uses for protocols are developed, these systems need to be modified to understand them to ensure that they are not producing false positives. Again, data mining proves useful for defining what constitutes normal use based on previous network activity.

4.4 Data Pre-processing

Feature selection is a fundamental part of the Data Mining Process. The main goal is to identify features that are important to the mining effort. The effort of feature selection is to reduce the dimensionality of the data to make processing the data more efficient. Within the study of data mining there is a phenomenon called "the curse of dimensionality" in which all the dataset members appear isolated and unique from the others. According to Dartigue, Jang and Zeng, the areas to analyze for feature selection and extraction can be in [12]:

- Intrinsic features which exist in all network traffic such as protocol, port, destination server name, and requester IP address
- Time-based features which connect traffic from "same host" or "same service" which is valuable in identifying DoS and fast probing exploratory attacks
- Host-based traffic features include grouping connections based upon the same server destination to help to identify slower probing attacks
- Content-based features that are designed to consider long term asynchronous conversations between the target server/service and the attacker's software client. These can be characterized as being slow, methodical and thorough attacks over wide windows in time

4.5 Model Development

Various data mining techniques have been explored in existing research to create Intrusion Detection Systems. Tsai, Tsu et al. performed a survey of machine learning techniques for intrusion detection seen in research papers between 2000 and 2007 [13]. Much of the research utilized training data to create classifiers which map input data to an output (benign or an intrusion). New incoming network traffic would be put through this classifier to determine if it represents an intrusion or not. The classifiers were generally one of three types: single, hybrid, or ensemble. Single classifiers utilize one single machine learning algorithm to create a single model which is used to make classifications. The most common single classifiers used to create IDPSs in the research are as follows:

- K-nearest neighbor (KNN) [14, 15]: instance based learning to classify a new vector based upon it's calculated nearest neighbor from the training set
- Support vector machines (SVM) [16]: a supervised model defining the decision boundary, gap between the most divergent training examples, based upon support vectors rather than the whole training set to classify new events
- Artificial neural networks (ANN) [17]: information processing units intended to mimic the network of neurons in the human brain for performing pattern recognition
- Self-organizing maps (SOM) [18]: an artificial neural network that uses unsupervised training to produce discretized representation of the training data in the form of a low-dimensional map
- Decision trees [15, 19]: maps feature observations about an event to conclusions learned from the features of a training dataset in the form of a classification/regression tree
- Naïve Bayes network [20]: analyzing the features independently of each other along a normal distribution as established by the training dataset
- Genetic algorithms [21]: a meta-heuristic designed to mimic natural selection in finding the most effective classification of new events based upon the features trained from the training dataset
- Fuzzy logic [22]: based upon a real world concept that things are never just black and white, rather they are in the spectrum of grey between the two extremes. It treats the training data as more benign and compares new data to be processed as more or less benign in comparison to the training set.

Hybrid classifiers combine multiple machine learning techniques to improve performance. This approach represents a more customized implementation to suit specific intrusion detection objectives. Hybrid classifiers may include multiple levels of processing/filtering of the training data where later phases are fed subsets of results from earlier filtering [23].

Ensemble classifiers are another effort to improve on single classifiers. They apply a collection (ensemble) of learning algorithms to different training samples to collectively provide improved performance [24].

	Integrity	*Availability*
Causative:		
Targeted	*The intrusion foretold*: mis-train a particular intrusion	*The rogue IDS*: mis-train IDS to block certain traffic
Indiscriminate	*The intrusion foretold*: mis-train any of several intrusions	*The rogue IDS*: mis-train IDS to broadly block traffic
Exploratory:		
Targeted	*The shifty intruder*: obfuscate a chosen intrusion	*The mistaken identity*: censor a particular host
Indiscriminate	*The shifty intruder*: obfuscate any intrusion	*The mistaken identity*: interfere with traffic generally

Fig. 4.2 Taxonomy of attacks against IDPS [8]

As data mining and machine learning tools become more popularly utilized methods for intrusion detection, they also become popular targets for adversaries to attempt to undermine. In computing, a denial-of-service (DoS) or distributed denial-of-service (DDoS) attack is an attempt to make a machine or network resource unavailable to its intended users. One common method of this attack involves saturating the target machine with external communications requests, so much so that it cannot respond to legitimate traffic or it responds so slowly it is rendered essentially unavailable. Such attacks usually lead to a server overload. For these types of attacks, the feature selection process becomes exceedingly more important. Computational resources can be optimized if critical features are detected and the noise is filtered away.

Barreno, Marco, et al. provide an excellent taxonomy of other approaches adversaries may use against typical IDPS [7]. These taxonomies are shown in Fig. 4.2.

According to the taxonomy, an attack is broken down into three different axes, influence, specificity, and security violation. The influence of an attack defines whether it is causative or exploratory. A causative attack modifies the training set that patterns are mined from in order to influence the learning model. An exploratory attack does not alter the training process, but rather uses other techniques to take advantage of existing weaknesses or blind-spots in the model. An attack is further classified by its specificity as being either targeted or indiscriminate. A targeted attack focuses on a specific intrusion or creating a specific misclassification while an indiscriminate attack looks for any possible intrusion. The third axis, security violation, focuses on the CIA (confidentiality, integrity, availability) model of a network by describing an attack as either an integrity attack or availability attack. An integrity attack results in the IDPS incorrectly classifying an intrusion as benign (false negatives) while an availability attack causes so many misclassifications (both false negatives and false positives) that the IDPS becomes unusable.

4.6 Malicious Code Detection

Within the scope of intrusion detection is the more specific security concern of malware or malicious code detection. As the prevalence of malware infections has reached epidemic proportions, it is becoming increasingly important to choose the right defenses to prevent costly malware infections that are targeted at stealing sensitive corporate secrets and mining critical user information records. With today's Internet, malware researchers are seeing a large spike in malware activity and estimate that thousands of new malware variants are being released into the wild daily. Working with large datasets and feature sets to discover hidden patterns has proved extremely applicable to the area of malware detection. Malware can be defined as a program that performs malicious behavior, compromises the security of the system, or performs a function against the wishes of the user. The spread of malware represents an increasing threat to maintaining the security of cyber systems. According to the Symantec Global Internet Security Threat Report, there were over 5,000 reported vulnerabilities in 2012 [25].

As mentioned in the previous section, traditional signature based detection is a standard approach for finding and detecting malicious behavior on a system. However, these methods are inherently less effective for detecting novel and polymorphic malware. Signature based detection cannot reliably detect new malware until after it has been identified and given a signature. Polymorphic malware attempts to continuously modify itself in order to evade detection from a previously assigned signature. These concepts pose a serious challenge to existing anti-virus solutions.

Automatic detection of malicious code is a common application of data mining techniques. One method for this detection is through the mining of auspicious binary executables. In order to perform this analysis, appropriate features must be selected to determine whether the sample is benign or malicious. These features may include a list of function calls, strings, headers, byte sequences, or other attributes of the binary [26]. These features can then be processed and fed into a classification algorithm. Some methods assign each sample a classification probability based on the Naive Bayes algorithm, a rules based classifier, or a multi-classifier system [26]. Oulette et al. proposed deep learning algorithms to classify related malware families using a more comprehensive understanding of the malware's intrinsic properties [27]. Others have developed solutions which extract n-gram features from both binary and assembly code [28].

Anon-trivial challenge of these approaches is finding and extracting relevant and useful features for the data mining. Another challenge of these approaches is that it can only classify new malware samples based on previous known samples. Also, various obfuscation techniques attempt to hide the true intent of the malicious code to skirt detection. In order to overcome these challenges, some solutions look for relevant features in a dynamic environment. These systems may search for anomalies within network traffic or other previously unseen behavior patterns. Thuraisingham et al. developed models using support vector machines to detect intrusions or malicious behavior based on deviations from normal network patterns [28]. In order to detect novel classes, Masud et al. proposed techniques for the detection of concept-drifts in data streams, which may be applied to the domains of network intrusion or fault detection [29]. These approaches must continually refine their techniques to gain acceptable detection and false positive rates. Since these detection methods are typically utilized with the oversight of a human analyst, a high false positive rate will quickly cause frustration for both the analysts and end users.

Although few commercial IDPS products currently utilize data mining, this is a topic of growing importance with a large (and growing) corpus of research supporting its use. As the number and complexity of existing exploits increases and it becomes easier and easier to morph and obfuscate attacks, most common IDPSs which rely on an updated database of known attack signatures will become less effective. Data mining techniques for learning generalized patterns indicative of attacks will soon become more prevalent and effective.

4.7 Data Mining for Improved Cyber Situational Awareness

Handling cyber threats unavoidably needs to deal with uncertainty and imprecise information. What is observed as potential malicious activities can seldom give us 100 % confidence on important questions about which machines have been compromised, the extent of damage that has been incurred, and who and why the systems have been targeted. It is through Social Network Analysis (SNA) that some of these questions may be answered. Again, this is a very complex problem which must take into consideration a wealth of information from multiple sources.

Efficient and reliable analysis of such large datasets is a challenge faced by both intelligence agencies and law enforcement. Data mining can yield results which would be impractical or impossible through manual efforts alone, due to the massive amount of relevant data available. These techniques are often performed semi-autonomously, delivering additional support for human analysts. Within the cyber security field, data mining processes may be applied in the defense of computer networks and cyber infrastructure to identify malicious actors or organizations that pose a threat. In addition, if some threatening entities have already been identified, then these techniques may be applied to expand the search in order to identify other related attackers.

Data mining provides the ability to correlate and condense data into a social network structure, in order to discover patterns and relationships between humans, organizations, or other entities. By representing a social network as a graph, with entities as nodes and relationships as edges, automated techniques can provide deeper insight into the social relationships present within that system. SNA techniques help the human analyst discover interesting factors or patterns that have previously unrecognizable. SNA provides mathematical constructs to model and predict useful patterns of social interactions. This analysis can greatly bolster the efforts of human analysts by identifying areas of interest, spotting emerging leaders, and predicting behavior. Krebs utilized SNA to identify core members of a terrorist network involved in the 9/11 attacks [906]. In this example, the relationships and structure were built from surveillance data released by government authorities and publicly available information on the web. This analysis discovered strong mutual connections between the hijackers, while also revealing an emerging leader within the network structure [30].

In addition to discovering individual entities within a social network, analysis can reveal the strength and influence of a network as a whole. Shang et al. developed an indicator model that measures the degree of connectivity of a network in order to find and predict criminal networks [31]. Iqbal et al. demonstrated the feasibility of the collecting online chat logs, identifying topics of conversation, and analyzing these messages for possible criminal activity [32]. Chen et al. developed techniques to identify strong subgroups within a network, and to find central members within a subgroup of a potential criminal network [33]. These data mining processes can provide key information in developing a clear understanding of the social dynamics in play within the social network.

In addition to passively understanding the social connections, this analysis can also provide direction for actively influencing the social network. This intelligence may help determine a course of action produce a desired effect within the organization. For example, if the key members of an organization can be identified, then crucial lines of communication may be intercepted or denied to alter the effectiveness of the group. Other techniques may be applied to relevant areas of the graph to achieve a certain desired effect.

This social network analysis often relies heavily on the mining of large datasets to construct these networks. Public social media sites are a common source for this data. Lau et al. produced mining methods which discovered both implicit and explicit relationships derived solely from public social media sites, through extracted words supplied to a probabilistic model [34]. While this analysis can be extremely powerful, it depends strongly on the quality of the data collected. If the data is biased, misrepresented, or incorrect, the results will similarly be erroneous.

4.8 Emerging Challenges for Data Mining in Cyber Operations

Modern and emerging networks are rated by the amount of billions of bits they can transport in a second, which uses the metric prefix of giga- to represent a one billion multiplication factor. A common rating of network bandwidth is the term Gigabit, and this rate is abbreviated as Gbps or Gb/s. In a single minute there can up to 60 Gigabits transferred, which is equivalent to 7.5 GB and is close to 1.5 DVDs worth of content. This number seems impressive at first, but quickly becomes shadowed when considering there are 1,440 min in a day, and the ratification of the IEEE standard 803.3ba defines both a 40 Gb/s and 100 Gb/s network [36]. In a single day, at a maximum sustained bandwidth of 100 Gb/s, over 219,142 DVDs worth of content could be transferred. These Internet bandwidth speeds are slowly moving to replace commercial infrastructure as the status quo. Google Fiber advertises it is 100 times the speed of broadband connections and is only at a bandwidth of 1 Gb/s [37]. However, there is a growing threat in cyberspace that will be able to block network traffic even at these high data rates

History was made in February 2014 when the largest ever Distributed Denial of Service (DDoS) attack was recorded by Cloudflare Incorporated [37]. Cloudflare is a content distribution network that hosts websites and applications for Internet users. The company recorded an attack of over 400 Gb/s against one of its hosted sites from a series of 4,529 vulnerable NTP servers [38]. Cloudflare also reports that Network Time Protocol (NTP) DDoS attacks see an amplification factor, of corrupted input to malicious-amplified output, of over 200 times, and they have observed the Simple Network Management Protocol (SNMP) protocol to have DDoS attacks with an amplification factor of 650 times [39]. Even with the worlds expanding Internet infrastructure of Fiber technologies these DDoS attacks will be able to saturate a company's Internet bandwidth, since they have currently shown an attack capability 4 times greater than the maximum 100 Gb/s bandwidth. Considering Google Fiber's speed claims: the NTP DDoS is 400,000 times greater in size than modern broadband cable bandwidth. Ultimately, data mining will be center stage in defense of growing DDoS and other unknown capabilities, since this focus is on massive amounts of data and bandwidth.

Analysts not leveraging data mining would become instantaneously saturated in extremely large data sets if they were to experience an NTP DDoS attack. Sifting through information that was being transmitted in a 1 Gb/s connection, or higher speeds, would need data mining to determine what activities and actions are occurring within this space. Data mining would allow the detection, determination, and prevention of cyber threats, which would enable IDPS to mitigate or even thwart such an attack. Any interesting scenario would be if an attacker was able to combine a DDoS with the execution of malicious code. The DDoS would then be used to obfuscate this malicious activity, and without data mining capabilities there could be a large delay time in discovering this code if it occurred at all. Future data mining will need to be able to be optimized in order to mitigate these near future cyber threat, and without leveraging data mining there seems to be no other solutions that would be able to allow maneuverability during an attack.

Maneuverability is key to cyber operations for both parties in a conflict. A DDoS attack is designed to effectively remove any movement of the target. Data mining could provide mitigation strategies that would allow a target partial survivability, which would allow transmission or even migration of operations to a non-attacked platform. Having extended periods of blocked transmissions in cyberspace could greatly cripple a system or asset with respect to denial, disruption, degrading, and even deception. A Key feature of future defenses must be able to survive and mitigate an attack to prevent a full stop of cyber maneuverability.

Furthermore new and different areas should start considering the application of data mining to potential big data problems. According to Kamal and Muccio (2011) mission awareness is at the heart of cyber situational awareness, which gives an understanding of mission to asset dependencies. In light of these new threats, and having this goal of situational awareness, new systems must incorporate data mining to stay relevant when analyzing big data. Having the ability to understand and interpret data through data mining will enable the ability to predict and provide potential courses of actions to defense systems. Lastly, there are applications for data mining in current and future cyber modeling, simulation, and war gaming. From participating in these events it has been observed that many Department of Defense war games rely heavily upon analyst input and interpretation of data. The addition of data mining in war games could provide a deeper analysis of the results, or even add the potential of multiple iterations of scenarios where currently there are only a few iterations. The application of data mining to cyberspace is endless, but it provides a greatly exciting future to all of those involved.

References

1. Jabbour, Kamal, and Sarah Muccio. "The Science of Mission Assurance." Journal of Strategic Security 4.2 (2011).
2. Han, Jiawei, Micheline Kamber, and Jian Pei. Data mining: concepts and techniques. Morgan kaufmann, 2006.
3. Applegate, Scott D. "The principle of maneuver in cyber operations." Cyber Conflict (CYCON), 2012 4th International Conference on. IEEE, 2012.
4. Gregorio-de Souza, Ian, et al. "Detection of complex cyber attacks." Defense and Security Symposium. International Society for Optics and Photonics, 2006.
5. Grant, Tim, Ivan Burke, and Renier van Heerden. "Comparing Models of Offensive Cyber Operations." Proceedings of the 7th International Conference on Information Warfare and Security: Iciw 2012. Academic Conferences Limited, 2012.
6. Villars, Richard L., Carl W. Olofson, and Matthew Eastwood. "Big data: What it is and why you should care." White Paper, IDC (2011).
7. Barreno, Marco, et al. "Can machine learning be secure?." Proceedings of the 2006 ACM Symposium on Information, computer and communications security. ACM, 2006.
8. Barreno, Marco, et al. "The security of machine learning." Machine Learning 81.2 (2010): 121-148.
9. Sabahi, F., and A. Movaghar. "Intrusion detection: A survey." Systems and Networks Communications, 2008. ICSNC'08. 3rd International Conference on. IEEE, 2008.
10. Scarfone, Karen, and Peter Mell. "Guide to intrusion detection and prevention systems (idps)." NIST Special Publication 800.2007 (2007): 94.
11. Han, Hong, Xin-Liang Lu, and Li-Yong Ren. "Using data mining to discover signatures in network-based intrusion detection." Machine Learning and Cybernetics, 2002. Proceedings. 2002 International Conference on. Vol. 1. IEEE, 2002.
12. Lee, W., & Stolfo, S. J. (1998). Data Mining Approaches for Intrusion Detection. Proceedings of the 7th USENIX Security Symposium. San Antonio.
13. Tsai, Chih-Fong, et al. "Intrusion detection by machine learning: A review." Expert Systems with Applications 36.10 (2009): 11994-12000.

14. Bishop, C. M. (1995). Neural networks for pattern recognition. England: Oxford University.
15. Mitchell, T. (1997). Machine learning. New york: McGraw Hill.
16. Vapnik, V. (1998). Statistical learning theory. New York: John Wiley.
17. Haykin, S. (1999). Neural networks: A comprehensive foundation (2nd ed.). New Jersey: Prentice Hall.
18. Kohonen, T. (1982). Self-organized formation of topologically correct feature maps. Biological Cybernetics, 43, 59–69.
19. Breiman, L., Friedman, J. H., Olshen, R. A., & Stone, P. J. (1984). Classification and regressing trees. California: Wadsworth International Group.
20. Pearl, Judea. (1988). Probabilistic reasoning in intelligent systems. Morgan Kaufmann
21. Koza, J. R. (1992). Genetic programming: On the programming of computers by means of natural selection. Massachusetts: MIT.
22. Zimmermann, H. (2001). Fuzzy set theory and its applications. Kluwer Academic Publishers.
23. Jang, J.-S., Sun, C.-T., & Mizutani, E. (1996). Neuro-fuzzy and soft computing: A computational approach to learning and machine intelligence. New Jersey: Prentice Hall
24. Kittler, J., Hatef, M., Duin, R. P. W., & Matas, J. (1998). On combining classifiers. IEEE Transactions on Pattern Analysis and Machine Intelligence, 20(3), 226–239.
25. Symantec. 2013 Internet Security Threat Report. Volume 18 Vol., 2013. Print.
26. Schultz, M. G., et al. "Data Mining Methods for Detection of New Malicious Executables". Security and Privacy, 2001. S&P 2001. Proceedings. 2001 IEEE Symposium on. Web.
27. Ouellette, J., A. Pfeffer, and A. Lakhotia. "Countering Malware Evolution using Cloud-Based Learning". Malicious and Unwanted Software: "The Americas" (MALWARE), 2013 8th International Conference on. Web.
28. Thuraisingham, B. "Data Mining for Malicious Code Detection and Security Applications". Intelligence and Security Informatics Conference (EISIC), 2011 European. Web.
29. Masud, M. M., et al. "Classification and Novel Class Detection in Concept-Drifting Data Streams Under Time Constraints." Knowledge and Data Engineering, IEEE Transactions on 23.6 (2011): 859-74. Web.
30. Krebs, Valdis E. "Mapping networks of terrorist cells." Connections 24.3 (2002): 43-52.
31. Xufeng Shang, and Yubo Yuan. "Social Network Analysis in Multiple Social Networks Data for Criminal Group Discovery". Cyber-Enabled Distributed Computing and Knowledge Discovery (CyberC), 2012 International Conference on. Web.
32. Iqbal, F., B. C. M. Fung, and M. Debbabi. "Mining Criminal Networks from Chat Log". Web Intelligence and Intelligent Agent Technology (WI-IAT), 2012 IEEE/WIC/ACM International Conferences on. Web.
33. Chen, Hsinchun, et al. "Crime data mining: an overview and case studies." Proceedings of the 2003 annual national conference on Digital government research. Digital Government Society of North America, 2003.
34. Lau, R. Y. K., Yunqing Xia, and Yunming Ye. "A Probabilistic Generative Model for Mining Cybercriminal Networks from Online Social Media." Computational Intelligence Magazine, IEEE 9.1 (2014): 31-43. Web.
35. Fayyad, U., Piatesky-Shapiro, G., Smyth, P., and Uthurusamy, R. (Eds.), 1996. Advances in Knowledge Discovery and Data Mining, AAAI Press, Cambridge
36. McCabe, Karen. "IEEE-SA - IEEE Launches Next Generation of High-Rate Ethernet with New IEEE 802.3ba Standard." IEEE Standards Association. Institute of Electrical and Electronics Engineers Standards Association, 26 May 2010. Web. 21 Feb 2014. https://standards.ieee.org/news/2010/ratification8023ba.html.
37. Prince, Matthew. "Technical Details Behind a 400Gbps NTP Amplification DDoS Attack.". Cloudflare, Inc, 13 Feb 2014. Web. 21 Feb 2014. http://blog.cloudflare.com/technical-details-behind-a-400gbps-ntp-amplification-ddos-attack.
38. Graham-Cumming, John. "Understanding and mitigating NTP-based DDoS attacks." . Cloudflare, Inc, 9 Jan 2014. Web. 21 Feb 2014. http://blog.cloudflare.com/understanding-and-mitigating-ntp-based-ddos-attacks.
39. Google Fiber Inc. "Plans and Pricing.". 2014. Web. 21 Feb 2014. https://fiber.google.com/cities/kansascity/plans.

Chapter 5
Trusted Computation Through Biologically Inspired Processes

Gustave W. Anderson

5.1 Introduction

In today's computing environments, one must assume that a subset of the system is currently, or will eventually be compromised. The proposed architecture supports design separation for high reliability and information assurance. By leveraging a hybrid fault model with multiple, parallel execution paths and resultant execution trace comparison, the proposed cognitive trust architecture identifies suspect nodes and assures trusted execution. Furthermore, the modeled architecture may be scaled through proactive thread diversity for additional assurance during threat escalation. The solution provides dynamic protection through distributing critical information across federated cloud resources that adopt a metamorphic topology, redundant execution, and the ability to break command and control of malicious agents.

5.1.1 Biologically Inspired Approach

The cognitive trust approach seeks to ensure continued operation under the assumption of imminent compromise and is modeled after the human immune system. The immune system is a system of biological structures and processes that protects against disease. For the immune system to function properly it must first detect the presence of a wide variety of agents. Conversely to compromise the system, pathogens must rapidly evolve and adapt to avoid detection and neutralization by the

G.W. Anderson (✉)
MacAulay-Brown, Inc. (MacB), 1 Riverside Circle, Suite 400A, Roanoke, VA, USA
e-mail: gustave7@gmail.com

R.E. Pino et al. (eds.), *Cybersecurity Systems for Human Cognition Augmentation*, Advances in Information Security 61,
DOI 10.1007/978-3-319-10374-7_5

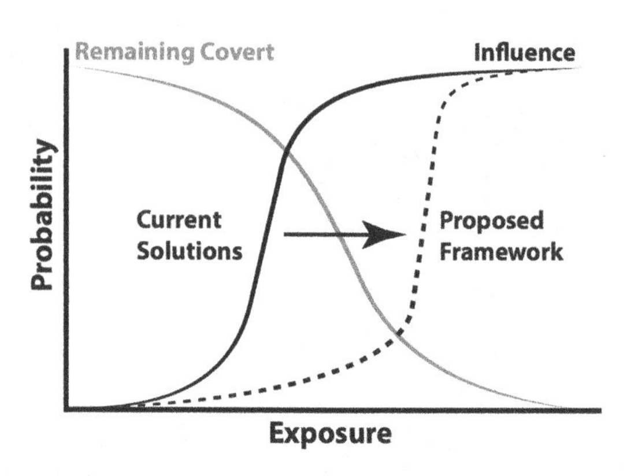

Fig. 5.1 POSM concept

immune system. To rapidly evolve, pathogens must gather information about the systems that they are trying to compromise. The more genetic information required for compromise; the increased risk of exposure.

Figure 5.1 represents the conceptual difficulty of compromising the proposed system. This figure demonstrates the conceptual likelihood of an attack being discovered (ordinate) as a function of the required exposure to the systems (abscissa) to achieve a successful attack. Traditionally, attacks need only to be as sophisticated as the weakest defense mechanism. While defense mechanisms must be as sophisticated as any possible class, type or quantity of attack. This imbalance yields a great advantage for the red forces. Currently, many systems require minimal system exposure in order to produce an effect to the system. Thus, if an adversary discovers an exploitable aspect of the system, they may achieve maximal effect while remaining covert. We propose that through a system of redundant execution paths and dynamic reconfiguration the threshold of system expose required for successful influence to a point that an attack may not launched successfully without exposing the presence of that attack.

5.1.2 Critical Assets

Currently most security solutions attempt to find the mythical "silver bullet" and achieve the "perfectly secure" system. In the end, there is no such thing. Next generation security systems must be mindful of the critical assets regardless if those assets are data or systems. With knowledge of the critical assets, the security system should then spend the vast majority of its effort detecting and protecting against attacks that attempt to compromise the critical assets. We assert that only then a "cost of compromise" vs "cost of prevention" analysis may be profitable.

In this chapter, we consider modern distributed computation environments, such as MapReduce [1] algorithms executed in a cloud-based system. In this scenario, we define the critical assets as (1) the IP of the algorithm and (2) the unimpaired execution of that algorithm. As a result, we spend minimal effort trying to prevent the compromise of nodes in our network; rather, we assume that a subset of the system is currently owned by an adversary. Consequently, our approach, like the immune system, is to make it mathematically infeasible for an adversary to simultaneously remain covert and exploit our critical assets.

5.2 Motivation

As the saying goes "You don't know where you're going unless you know where you've been." Regarding the protection of modern computing systems, it is beneficial to take a step back and consider computing security from a historical perspective [2]. The 1970s represented "the age of the Mainframe". Advances in large memory devices allowed for the processing of large amounts of data on a thick back-end while users operated on a thin-client. This operating paradigm brought to light new operating system and information security issues present with multi-user systems. Some seminal work responding to these challenges include: Saltzer and Schroeder [3], data encryption standard (DES) [4], Diffie-Hellman [5] and RSA [6]. Miniaturization of computing hardware in 1980s brought about "the age of the Personal Computer (PC)." With computers becoming small enough to fit under a desk, there was no longer a need for room-sized computers. With this shift in computation multi-user and multi-level security became largely irrelevant, exposing a whole new set of challenges. Separation of duties concepts previously developed for multi-level security inspired several advancements in computing security during this decade including the Clark Wilson [7] and Chinese wall [8] models. With the advent of hypertext transfer protocol (HTTP), hypertext markup language (HTML) and graphical web browsers the 1990s brought about "the age of the Internet." Again this new operating environment exposed new threats. The Morris worm in 1988, representing the beginning of Internet worms observed in the wild. In addition, Alephs' seminal publication "Smashing the Stack for Fun and Profit, [9]" demonstrated the capability to remotely execute unauthorized remote code. The security community responded with a variety of network security architectures yielding the creation of firewalls and intrusion detection systems. The growing popularity of the Internet brought about "the age of the Web" in the 2000s. Server-side and client side scripting languages provided for a dynamic web experiences, which resulted in nearly every household having a personal computers and Internet access. This growth in user base paved the way for the electronic commerce industry. This new operating environment, again, exposed new threats include SQL injection, cross-site scripting and attacks against the domain name system. The growth of wireless communication methods (both 802.11 and cellular) provided ubiquitous communication and lead to "the modern age of mobile computing." Truly history has repeated itself; modern computation is often performed on smartphones and tablet PCs (thin-client) with data stored "in the cloud" (thick back-end).

5.2.1 Known Attack Learned Defense

The security community has shown tremendous adaptability over the years. For each new operating paradigm, new techniques and methods were developed to protect against malicious behavior. Historically, the community has built solid defenses but as the paradigm shifts so has the attacks surface and we again need to adjust our posture to defend against the new potential attack vectors. Unfortunately, going forward this reactionary approach is ultimately unsustainable. Modern systems are increasingly complex, yielding larger attack surfaces and technologies are evolving at an increasing rate. Furthermore, the general availability of communication allows for an attack to be launched from anywhere at anytime by anyone. We assert that rather than protecting against specific classes of attacks it is beneficial to focus protection mechanisms on making it infeasible for an adversary to simultaneously exploit the critical assets of a system while remaining convert.

5.3 POSM Approach

The proposed framework forces an adversary desiring to influence a cloud based system to modify such a large part of the system that it is infeasible to remain covert while attempting to do so. The POSM approach outlines the goals of the framework, namely: (1) Protection of Intellectual Property, (2) Operation under attack, (3) Separation of power and (4) Mutation of configuration.

The POSM framework protects critical operations through a system built on separation of power. That is to separate: (1) the responsibility of algorithm or application execution and (2) data consolidation. Execution of application and algorithms are distributed throughout the system in separate and distinct pieces. In addition, the control of the system will be distributed to separate nodes. Our system is separated into two classes of application sets. Figure 5.2 demonstrates the two classes of nodes that will exist in the system, namely scheduling nodes (blue with "sched" text) and task nodes (gray-green with "node" text). Scheduling nodes are tasked with the distributed control along with the tracking and execution of the application. While task nodes are provided with independent slices of the algorithm and tasked with the execution of these slices. Task nodes are not provided contextual knowledge pertaining to which slice of the algorithm they have nor what the input they are provided represents. Scheduling nodes are aware of how many nodes exist, and which nodes contain the algorithm slices, yet are unaware of the specifics of the algorithm slice execution.

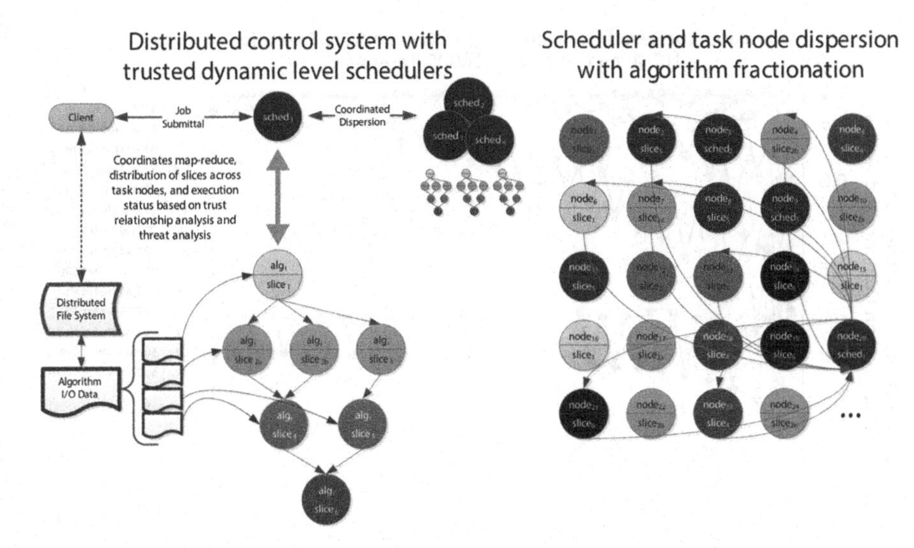

Fig. 5.2 POSM scheduling and task node operation

5.3.1 Redundant Execution

Increased redundancy provides resiliency against influence by forcing an adversary to compromise a greater number of nodes in the system. Furthermore, by increasing the quantity of nodes an adversary must compromise we naturally increases the exposure necessary, and thus increase the likelihood that their behaviors may be observed (we discuss this in more detail in Sect. 5.3.2). When considering the level of redundancy it is beneficial to more generally consider nodes as "good" or "bad", rather than specifically classifying them as "compromised" or "not compromised". For the purpose of protecting against incorrect computation results, it does not matter if these incorrect values were due to adversarial influence or failing/misconfigured hardware. To calculate the level of redundancy we define models for development and verification. We assume that we may obtain, via calculation or specification, the amount of redundancy necessary for a given random distribution of "good" and "bad" nodes required to achieve a defined probability of selecting more "good" than "bad" nodes.

5.3.1.1 Comparison of Individual Executions

The first model we define and evaluate is the comparison of individual slice execution. This model defines a probability P that represents the chance for a given slice that more slices will execute on "good" nodes than on "bad" nodes.

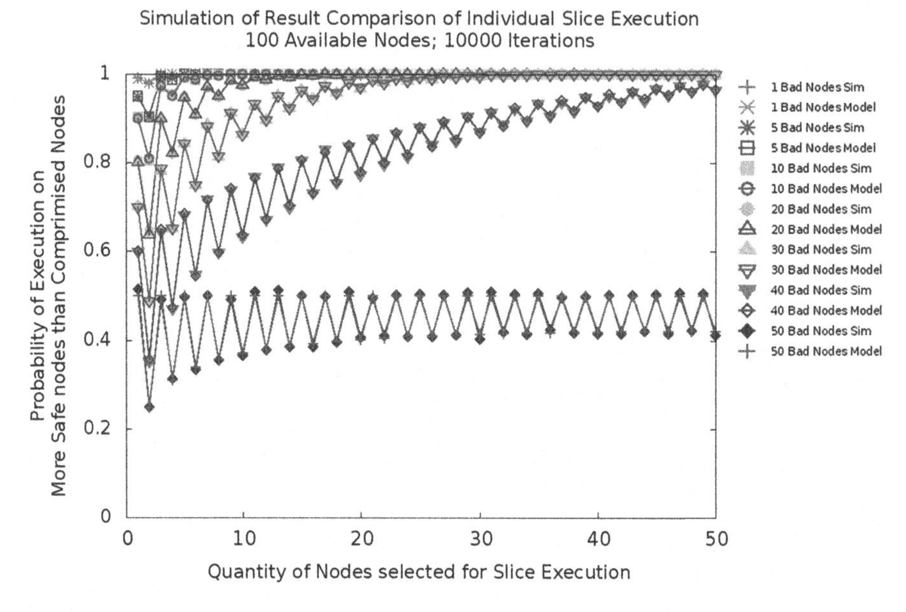

Fig. 5.3 Verification of model 1

$$P = \sum_{i=\left\lfloor \frac{n}{2} \right\rfloor + 1}^{n} \frac{\binom{m}{i}\binom{N-m}{n-i}}{\binom{N}{n}} \tag{5.1}$$

Where N represents the number of nodes available, m represents the number of good nodes, N—m the number of suspect/compromised nodes and n is the number of nodes selected for redundant slice executions. This provides the control system guidance in determining how much redundancy to introduce to meet a desired trust metric.

Figure 5.3 results from an Octave [10] simulation where a number of nodes (abscissa) are pseudo-randomly selected from a population of 100 available nodes for parallel execution. Of these 100 nodes some are pseudo-randomly selected as "Bad" (demonstrated by the varying colored lines). The ordinate demonstrates the probability that more of the select nodes are considered "good" than are considered "bad." As we can see the model derived in Eq. 5.1 very closely follows what happens in the simulation. In addition, we see that increasing the redundancy increases the probability of obtaining more safe nodes, and thus result comparison will yield an unimpaired execution. The saw-tooth nature of the plot results from the fact that it is always more beneficial to compare odd numbers of nodes as with an even quantity of nodes it is possible to have the same number of "good" and "bad" nodes. Furthermore, in the extreme scenario where half of the network is compromised we can see that it is unlikely we will be able to achieve a scheme that will yield a beneficial result. However, if the adversary is able to influence more than half of the system, it is unlikely any scheme will survive an attack of this magnitude. Finally, we can see that we have verified that this distribution very closely follows the behavior of hyper-geometric distribution. Among other reasons this conclusion is important, is that it allows us to consider probabilities without replacement if N and m are sufficiently large in comparison to n.

5.3.1.2 Comparison of the Result of Parallel Threads

Extending the previous model, we define a model that calculates the probability of successful comparison of parallel execution threads. Equation 5.1 describes the probability of successful comparison of each individual slice for each step in the algorithm. If we implement algorithm diversity, it may not be possible to compare the result of individual steps in the algorithm; thus we need a model that is useful for performing only result comparison of the entire thread across p nodes.

$$P = \sum_{i=\left\lfloor \frac{q}{2} \right\rfloor +1}^{q} \left[\frac{q!}{i!(q-i)!} \prod_{j=1}^{i} \frac{\frac{m-(j-1)p}{p}}{\frac{N-(j-1)p}{p}} \prod_{k=i+1}^{q} \left(1 - \frac{\left(\frac{m-i(k-i-1)p\frac{m-ip}{N-ip}}{p} \right)}{\left(\frac{N-(k-1)p}{p} \right)} \right) \right] \tag{5.2}$$

Equation 5.2 demonstrates that probability P of a successful comparison of redundant thread execution. Where N represents the number of nodes available, m represents the number of good nodes, N—m the number of suspect/compromised nodes, p is the number of nodes comprising a thread execution and q is the number of threads selected for redundant executions. This model may also be used to provide the control system guidance in determining how much redundancy to introduce to meet a desired trust metric.

5.3.2 *Cognitive Trust*

Though redundant execution allows for resiliency against adversarial influence, redundant execution alone will not stop a resolute adversary. The POSM framework is built on a cognitive trust model where, like a social network, each element in the system observes how other elements behave and builds a trust value accordingly. This concept of trust allows the framework to then favor elements that are behaving more trust-worthily in lieu of elements that are behaving abnormally.

We have extended previous cognitive trust models [11] by drawing from neuro-scientific approaches of how humans determine and generate trusted relationships with one another. Psychologists have widely defined trust as:

> *Trust is a psychological state comprising the intention to accept vulnerability based upon positive expectations of the intentions or behavior of another* [12].

That is to say, one who trusts another willingly exposes themselves to risk. Mayer, et al. [13] categorized these trust evaluations into three categories: ability, integrity and benevolence. Depending on how one perceives another across these three categories will determine how much risk they are comfortable exposing themselves to.

Ability refers to how one assesses another person's knowledge competency or skill to complete a given task. Integrity is the degree in which one adheres to accepted principles; this is based on consistency of past actions, credibility of communication, commitment to standards of fairness, and the congruence of the others word and deed. Finally, Benevolence refers to ones assessment of how concerned an individual is about their welfare to either advance their interests, or at least not impede them; in many ways, this may be considered the emotional response [13].

The level of trust evolves as parties interact. Positive interactions yield increasing levels of trust while negative interactions erode trust. The challenge for both evaluating trust in the human and computing contexts is quantizing behaviors and interactions into a numeric value that accurately reflects how much one should / does trust another. Through simulation, we demonstrate an enumeration of trust as a function of previous successful and unsuccessful interactions. This trust evaluation was then used to weigh redundant executions for the sake of favoring elements that are behaving more trust-worthily in lieu of elements that are behaving abnormally.

Figure 5.4 demonstrates a baseline strategy of a redundant execution of a MapReduce computation. Under this strategy, after completion of the computation task (performed by task nodes) and during the reduce task (performed by scheduling nodes) each scheduling node shares with the other scheduling nodes the results provided by each of their tasks nodes. Then individually the scheduling nodes define the "correct" result as the majority result. This simulation demonstrates again that increased redundancy does in fact reduce the chance a compromised node will affect the result. In addition, the saw-tooth nature of the plot results from the fact that it is always more beneficial to compare odd numbers of nodes. With an even quantity of nodes it is possible to have the same number of "good" and "bad" nodes and the correct final result is unclear. Finally, with this strategy it is unlikely that execution will conclude with the correct result if more than 20 % of the population is compromised.

Figure 5.5 demonstrates an extension of the previous strategy by weighing the reported result with the trust evaluation of task node that provided the result. The trust evaluation is derived from the combination of ability, integrity and benevolence metrics. Like the pure strategy, we see that increased redundancy does in fact reduce the chance a compromised node will affect the result. However, with this strategy the probability that a compromised node will affect the final result is significantly reduced. In addition, we see a reduction in the saw-tooth nature of the

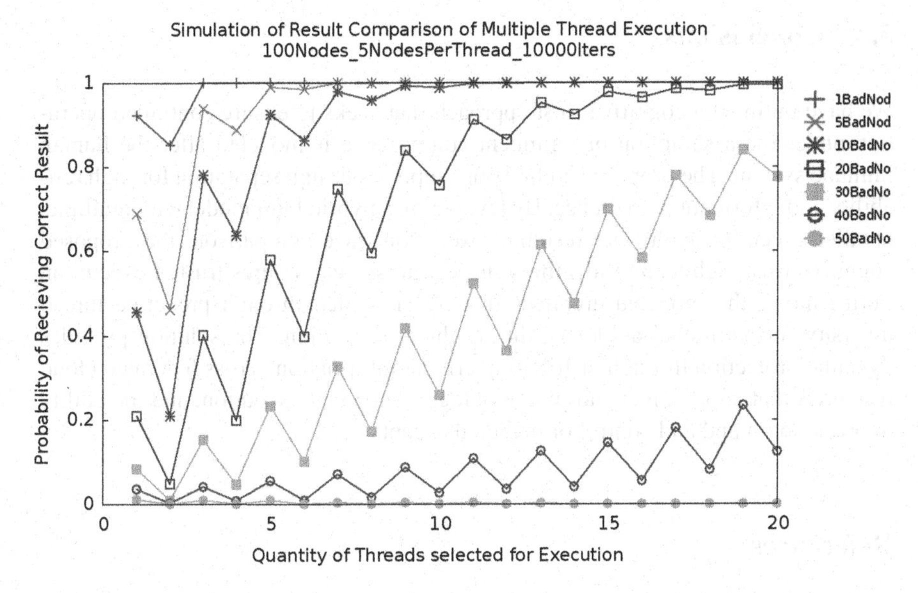

Fig. 5.4 Pure strategy

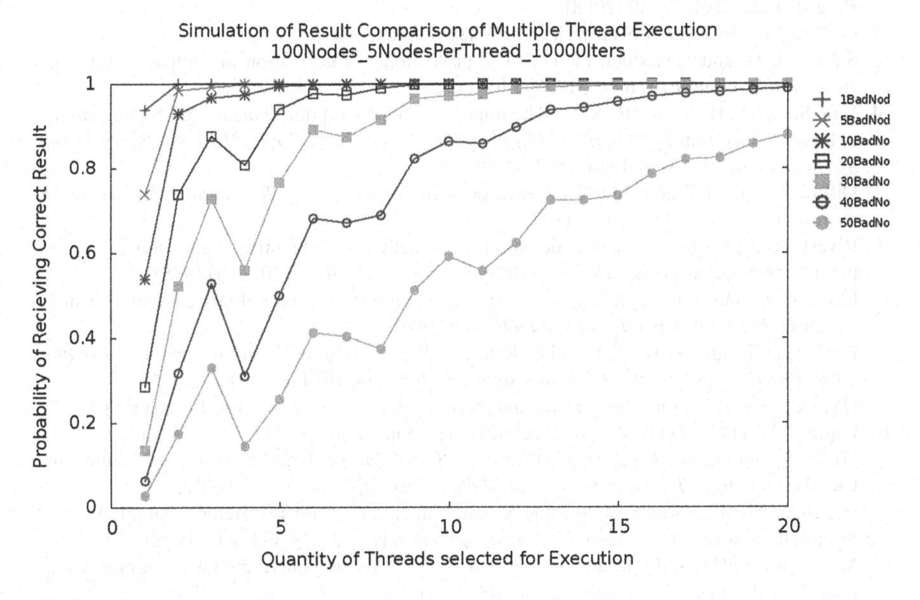

Fig. 5.5 Trust augmented strategy

plot. This is due to the fact that even when the same number of good and bad nodes are selected for a given slice, the trust evaluation allows for a tie break; yielding a higher likelihood of achieving the correct result. Finally, the dips observed at 4, 8 and 11 threads result from the adversarial advantage of comprising a scheduling node. Under these simulation parameters it is much easier for an adversary to remain covert while still attempting to influence the final result.

5.4 Conclusions

We have outlined a cognitive trust approach that seeks to ensure continued operation under the assumption of imminent compromise is modeled after the human immune system. The proposed architecture supports design separation for high reliability and information assurance. By leveraging a hybrid fault model with multiple, parallel execution paths and resultant execution trace comparison, the proposed cognitive trust architecture identifies suspect nodes and assures trusted execution. Furthermore, the modeled architecture may be scaled through proactive thread diversity for additional assurance during threat escalation. The solution provides dynamic protection through distributing critical information across federated cloud resources that adopt a metamorphic topology, redundant execution, and the ability to break command and control of malicious agents.

References

1. Lammel, R., "Googles MapReduce programming model Revisited," *Science of Computer Programming* 70(1), 1–30 (2008).
2. Gollman, D., [*Computer Security*], John Wiley & Sons (2011).
3. Saltzer, J. H. and Schroeder, M. D., "The protection of information in computer systems," *Proceedings of the IEEE* 63(9), 1278–1308 (1975).
4. Han, S.-J., Oh, H.-S., and Park, J., "The improved data encryption standard (DES) algorithm," in [*Spread Spectrum Techniques and Applications Proceedings, 1996., IEEE 4th International Symposium on*], 3, 1310– 1314, IEEE (1996).
5. Diffie, W. and Hellman, M., "New directions in cryptography," *Information Theory, IEEE Transactions on* 22(6), 644–654 (1976).
6. Rivest, R. L., Shamir, A., and Adleman, L., "A method for obtaining digital signatures and public-key cryptosystems," *Communications of the ACM* 21(2), 120–126 (1978).
7. Clark, D. D. and Wilson, D. R., "A comparison of commercial and military computer security policies," *NIST SPECIAL PUBLICATION SP* (1989).
8. Brewer, D. F. and Nash, M. J., "The chinese wall security policy," in [*Security and Privacy, 1989. Proceedings., 1989 IEEE Symposium on*], 206–214, IEEE (1989).
9. One, A., "Smashing the stack for fun and profit," *Phrack magazine* 7(49), 14–16 (1996).
10. Eaton, J. W. et al., [*GNU octave*], Free Software Foundation (1997).
11. Wang, W., Zeng, G., Tang, D., and Yao, J., "Cloud-DLS: Dynamic trusted scheduling for Cloud computing," *Expert Systems with Applications* 39(3), 2321–2329 (2012).
12. Rousseau, D. M., Sitkin, S. B., Burt, R. S., and Camerer, C., "Not so different after all: A cross-discipline view of trust.," *Academy of management review* 23(3), 393–404 (1998).
13. Mayer, R. C., Davis, J. H., and Schoorman, F. D., "An integrative model of organizational trust," *Academy of management review*, 709–734 (1995).

Chapter 6
Dynamic Logic Machine Learning for Cybersecurity

Leonid Perlovsky and Olexander Shevchenko

6.1 Introduction

Today's networks and their users are under attack from an ever-expanding universe of threats and malware. Malware are malicious software codes that typically damages or disables, takes control of, or steals information from a computer system. Malware broadly includes botnets, viruses, worms, Trojan horses, logic bombs, rootkits, boot kits, backdoors, spyware, adware, and other types of threats. The ever increasing danger of the future threat is its ability to evolve for avoiding system defenses. Future threats will be using machine learning to outsmart the defenses. Therefore the future of cybersecurity if a warfare of machine learning techniques. The more capable machine learning technique will win.

Correspondingly, an important direction of cybersecurity concentrates on machine learning techniques [4, 7, 10, 21, 53]. In this Chapter we discuss machine learning techniques based on dynamic logic (DL), which can be mathematically proven to have the fastest possible learning ability [48]. Steps toward developing such cybersecurity methods are discussed below.

This chapter describes an adaptive machine learning techniques based on abstract models. The approach to detecting novel attacks is anomaly detection: we develop algorithms learning models of attack-free traffic, and then detect deviations identifying malware. Gradual learning is a fundamental aspect of this approach. We begin assuming that an adequate protection system exists, and we can learn characteristics-models of attack-free traffic. The developed algorithms learn evolution of the

L. Perlovsky (✉)
LP Information Technology & Harvard University, Cambridge, MA, USA
e-mail: lper@rcn.com

O. Shevchenko
LP Information Technology, Cambridge, MA, USA

R.E. Pino et al. (eds.), *Cybersecurity Systems for Human Cognition Augmentation*, Advances in Information Security 61,
DOI 10.1007/978-3-319-10374-7_6

malware as it attempts to hide its harmful nature. For the success of this approach, learning of the defensive system must be faster than evolution of the threat.

The defensive system learns to recognize threats as combinations of basic elements, words or n-grams. In principle, this is a most general and universal approach, potentially capable of recognizing any threat. The difficulty of realizing this universal potential is computational complexity and slow learning of most existing algorithms. The reason for these difficulties is fundamental: the number of combinations is very large, even relatively few n-grams can be used to form a very large number of combinations. The number of combinations of only 100 n-grams is 100^{100}, this number exceeds all interactions of all elementary particles in the Universe during its entire lifetime. Therefore even if the entire Universe could be made to learn combinations of n-grams, it will not be able to perform its job fast enough. Later we will relate this fundamental difficulty to Gödelian difficulties of logic.

We describe DL, a mathematical technique overcoming Gödelian limitations. DL has been used to overcome combinatorial complexity (CC), a difficulty that for 50 years has prevented classical pattern recognition and artificial intelligence to solve many complex problems, such as detection of patterns below noise and among unrelated signals (clutter). The developed DL algorithms have overcome CC and improved detection performance by orders of magnitude. After introducing DL we discuss CC specific to cybersecurity. This CC is related to learning *structures* of threat models. The past research in dynamic logic can be understood as developing continuous mathematical representation of associations between signals and models. The current overcoming of CC of learning cyber-security models requires continuous mathematical representation of model *structures*. Model structure is a combination of inherently discrete mathematical constructs; representing it continuously is equivalent to eliminating a difference between continuous and discrete mathematics in this wide field. This will overcome Gödelian limitations of classical logic in this field.

This requires the new mathematical method described here. We outline an approach to proving the fastest possible learning of DL. We illustrate the new DL technique of machine learning using an abstract simulated data set, and finally, we demonstrate DL using publicly available malware data bases.

In addition to being a provably fastest machine learning technique, DL is also an adequate model for several fundamental mechanisms of the mind. We briefly discuss theoretical foundations for cognitive-emotional functions of DL and their experimental proofs in several brain-imaging labs. The combination of mathematical and cognitive superiority promises reliable future cybersecurity.

6.2 CC of Machine Learning During the Last 50 Years and the Gödelian Problem in Logic

Developing machine learning techniques exceeding human learning abilities started in the 1950s, when computers become available. Since then, hundreds and thousands learning algorithms have been developed. Some decades were dominated by a particular types of techniques or paradigms [27]. Early neural networks including perceptrons were developed in the 1950s. The 1960s were dominated by rule-based artificial intelligence. The 1970s witnessed a resurgence of interest in neural networks. Meanwhile statistical pattern recognition and fuzzy logic have been developed. Many algorithms achieved success in an area for which they have been developed. Some algorithms seemed to have a capability to be generalized to much wider areas and to become a general paradigm of intelligence. Yet a general algorithmic approach exceeding human intelligence remained an unattainable goal. During the 1990s and 2000s machine learning has been incorporating "lessons learned" in psychology, attempting to model mechanisms of the human mind; yet psychology itself faced difficulties when attempting to model mechanisms of cognition. Many recent "cognitive algorithms," attempting to follow mechanisms of the mind studied in psychology, use the same mathematical methods that have been known for decades. They have been known to look like they can solve a variety of problems, but practically they have been limited to "toy" problems and could not be generalized to different or more complex problems. Gradually it became accepted that all previously used paradigms faced the difficulty of CC.

CC as a general problem encountered by all paradigms attempting to model the mind and to create equally capable machine learning techniques have been discussed in [25]. This publication analyzed all major paradigms of machine learning and for each identified a reason for CC. In parallel a fundamental reason for CC has been identified [24, 27]. Using today's understanding [46, 48] it can be formulated succinctly: CC is related to logic; it is a manifestation of difficulties of logic discovered by Gödel in the 1930s [11].

Arguments can be formulated as follows: learning patterns to be recognized requires storing in machine memory many variations of patterns resulting in complexity of learning. But much worse complexity is related to a need to find a pattern of interest among other patterns of no interest (clutter), which could partially overlap. This results in a need to learn all (or nearly all) combinations of pixels in an image (or samples in a signal) and associating them with appropriate patterns or clutter. The number of combinations is very large, as discussed it is practically infinite. A problem of limited complexity in terms of the number of patterns of interest, or even in terms of the number of pixels or samples, acquires practically infinite complexity, due to a need to consider combinations.

As discussed in given references, the argument summarized above closely follows Gödelian arguments that lead to fundamental difficulties in logic. Gödel [11] considered all logical statements, including potentially infinite ones. And he demonstrated that although he logically listed all logical statements, he was able to prove that there have to be logical statements not in the list. This Gödelian argument can be related to our argument above: the Gödelian list included *combinations* of the original elements forming logical statements. And although the number of original elements was infinite, the number of combinations turns out to be a "significantly larger" infinity. Whereas the original infinity was countable, the number of combinations turn out to be uncountable infinity. To prove a *fundamental* difficulty in logic, Gödelian argument has to be applied to a infinite system, such as logic. Applying the Gödelian argument to a finite system, such as computer, does not result in a "fundamentally irresolvable" difficulty. Instead it results in CC, a "practically fundamental" difficulty. Thus for the purpose of designing efficient machine learning algorithm, CC is as fundamental as Gödelian difficulty in logic.

Understanding of this fundamental reason for CC is essential for making progress in machine learning. Thousands of algorithms have been designed for machine learning since the 1950s. These attempts still continue. The argument above demonstrates that unless fundamental reliance on logic during algorithm design is avoided, CC will persist. But what does it mean to avoid logic? Isn't entire science based on logic? Should the entire science be abolished? What could be used instead? The mind solves problems that computers cannot solve. Young kids and even birds solve problems that computers cannot. Can we understand how minds do this? Can this understanding be scientifically formulated?

6.3 Dynamic Logic

For thousands of years logic has been the best way to conduct arguments, including scientific arguments. The Newtonian physics, quantum physics, theory of relativity are based on logic. Only recently, when facing problems related to working of the mind we encountered insufficiency of logic. It is interesting that Aristotle, the founder of logic, did not use logic when explaining working of the mind. To explain mind, Aristotle developed theory of forms. Aristotelian forms are different from Platonian ideas and from contemporary understanding of concepts of the mind. Ideas of Plato and concepts in contemporary psychology are static entities, similar to logical statements, such as: "this is a chair." Instead, Aristotelian forms are dynamic entities, processes in which "mind meets matter." Today we describe it as an interaction between top-down and bottom-up neural signals. Before Aristotelian forms meet matter they exist as *potentialities*; in interacting with matter they become *actualities* [1].

What this process-logic means mathematically, in which way is it fundamentally different from usual classical logic? And why after Gödel has proven that logic has fundamental irresolvable difficulties in the 1930s, computer scientists still attempt to develop machine learning using logic? Did not neural networks and fuzzy logic attempt to overcome difficulties of logic?

This section presents dynamic logic, a mathematical theory that closely follows a process-logic of Aristotelian forms. After presenting mathematical formulation we return to discussion of the above questions: why despite Aristotle and Gödel, machine learning has been developed using classical logic. We discuss that the reason has been no less fundamental than logic itself. But first let us present a mathematical formulation of dynamic logic, a process logic, following [27, 30].

DL maximizes a similarity L between the data $\mathbf{X}(n)$, n = 1… N, and models $\mathbf{M}(m)$, m = 1… M,

$$\mathrm{L} = \prod_{n \in N} \sum_{m \in M} \mathrm{r}(\mathrm{m})\,\ell(\mathrm{X}(\mathrm{n}) \mid \mathbf{M}(\mathrm{m})) \tag{6.1}$$

Here l (**X**(n) | **M**(m)) are conditional similarities, later we denote them l (n|m) for shortness; they can be defined so that under certain conditions they become the conditional likelihoods of data given the models, L becomes the total likelihood, and DL performs the maximum likelihood estimation. Coefficients r(m), model rates, define a relative proportion of data described by model m; they must satisfy a condition,

$$\sum_{m \in M} \mathrm{r}(\mathrm{m}) = 1. \tag{6.2}$$

A product over data index n does not assume that data are probabilistically independent, relationships among data are introduced through models. Models **M**(m) describe alternative states of the system; expression (1) accounts for all possible alternatives in the data through all possible combinations of data and models. Product over data index n of the sums of M models results in M^N items, which is the mathematical reason for CC.

Learning consists in estimating model parameters, which values are unknown and should be estimated along with r(m) in the process of learning. Among standard estimation approaches is the multiple hypothesis testing algorithm [54], which considers every item among M^N. Logically this corresponds to considering separately every alternative association between data and models and choosing the best possible association (maximizing the likelihood). It is known to encounter CC.

DL avoids this logical procedure and overcomes CC as follows. Instead of considering logical associations between data and models, DL introduces continuous (or soft) associations,

$$\mathrm{f}(\mathrm{m}|\mathrm{n}) = \mathrm{r}(\mathrm{m})\ell(\mathrm{n}|\mathrm{m}) / \sum_{m' \in M} \mathrm{r}(\mathrm{m}')\ell(\mathrm{n}|\mathrm{m}'). \tag{6.3}$$

The DL process for the estimation of model parameters $\mathbf{S}_m$ begins with arbitrary values of these unknown parameters with one restriction; parameter values should be defined so that partial similarities have large variances. These high-variance uncertain states of models, in which models correspond to any pattern in the data, correspond to the Aristotelian potentialities. In the process of estimation variances are reduced so that models correspond to actual patterns in data, Aristotelian actualities. This DL-Aristotelian process "from vague to crisp" is defined as follows:

$$\begin{aligned} &\mathrm{df}(\mathrm{m}|\mathrm{n})/\mathrm{dt} = \mathrm{f}(\mathrm{m}|\mathrm{n}) \sum_{m' \in M} \left[\delta_{\mathrm{mm}'} - \mathrm{f}(\mathrm{m}'|\mathrm{n})\right]\left[\partial \ln l(\mathrm{n}|\mathrm{m}')/\partial \mathbf{M}_{\mathrm{m}'}\right](\partial \mathbf{M}_{\mathrm{m}'}/\partial \mathbf{S}_{\mathrm{m}'})\mathrm{d}\mathbf{S}_{\mathrm{m}'}/\mathrm{dt}, \\ &\mathrm{d}\mathbf{S}_{\mathrm{m}}/\mathrm{dt} = \sum_{n \in N} \mathrm{f}(\mathrm{m}|\mathrm{n})\left[\partial \ln l(\mathrm{n}|\mathrm{m})/\partial \mathbf{M}_{\mathrm{m}}\right]\partial \mathbf{M}_{\mathrm{m}}/\partial \mathbf{S}_{\mathrm{m}}, \delta_{\mathrm{mm}'} = 1 \text{ if } \mathrm{m} = \mathrm{m}', 0 \text{ otherwise.} \end{aligned} \tag{6.4}$$

Parameter *t* here is an internal time of the DL process; in digital computer implementation it is proportional to an iteration number.

A question might come up why DL, an essentially continuous process, seemingly very different from logic is called logic. This topic is discussed from a mathematical viewpoint in [16, 56, 57]. Here I would add that DL explains how logic emerges in the mind from neural operations: vague and illogical DL states evolve in the DL process to logical (or nearly logical) states. Classical logic is (approximately) an end-state of the DL processes.

6.4 Illustration of DL Algorithms

Before discussing the development of DL algorithms for Cybersecurity and specific difficulty that this poses, in this section we illustrate DL for detecting a simple pattern in a difficult condition, below noise and clutter. The pattern in this case is a point object moving on a straight line. This problem is called tracking. It is among most important practically needed pattern recognition problems: it is required for military and civilian applications. Every airport has several tracking systems, which have to be perfect so that airplanes can be safely directed in any weather. Tracking algorithms have been developed since WWII. Improving these algorithms by 1 % requires significant research and would be often judged a good result for a Ph.D. thesis. The difficulty of the case considered here is that signals coming from the object are *significantly* below signals of no interest, clutter. The object signals in this example are approximately 70 times lower than has been required by the best previous tracking algorithms (the required improvement thus is 7,000 %). Problems of such difficulty have been considered unsolvable; DL makes them possible to solve [48].

We consider two-dimensional data $\mathbf{X}_n$, vectors in (x,y). These data are illustrated in Fig. 6.1b. This figure shows N = 3,018 data points in (x,y) for 6 scans at time $N = T_n$, n = 1,... 6; all 6 scans are shown here superimposed. These 3,018 data points include 3,000 clutter signals and 18 signals coming from 3 moving objects, 1 signal from each object in each scan. Target signals are on average lower than clutter signals, therefore they cannot be differentiated from clutter in this figure. The clutter model, m = 1, is a uniform model normalized on the size of the data scan, length(x)*length(y); its only free parameter is the model rate, characterizing the proportion of clutter:

$$\ell(n|m=1) = 1/\text{length}(x) * \text{length}(y); r(1); \tag{6.5}$$

The total number of models, M includes the number of objects, M-1, which is unknown. Therefore r(1) is unknown as well. Objects over a short period of time move over straight lines. In this case an object model is

$$\text{for } m=1,\ \mathbf{M}(m) = \mathbf{MX}(m) + \mathbf{MV}(m) * T_n; \text{and } r(m) = 6/N. \tag{6.6}$$

here **MX**(m) is an unknown initial position of object m, **MV**(m) is an unknown velocity of object m, these are two-dimensional vectors in (x,y), and T_n is a known time of each n-th data point, counting from the initial scan. Conditional similarity for model m = 1 is a Gaussian function

$$\ell(n|m) = (1/2\pi v)\exp\left(-\left(\mathbf{X}(n) - \mathbf{M}(m)\right)^2 / 2v\right). \tag{6.7}$$

Here v is variance of sensor errors in (x,y). DL process estimating model parameters (r(m), **MX**(m), **MV**(m)) is given by Eq. 6.4, using Eqs. 6.5–6.7. To initiate the DL process, object positions and velocities are taken arbitrary, variance is taken large, equal the size of Fig.6.1; the initial number of objects is taken 1 (M = 2). After every step in DL Eq. 6.4, an algorithm attempts to increase the number of models by 1 (or decrease by 1, when it is more than 2), and keeps M corresponding to a larger likelihood value. The results are illustrated in Fig.6.1.

Three moving objects without clutter are shown in Fig. 6.1a. The data as measured with object signals and clutter are shown in (b); there are 6 radar scans, each scan showing 3 object signals and 500 clutter signals, total of 3,018 data points;

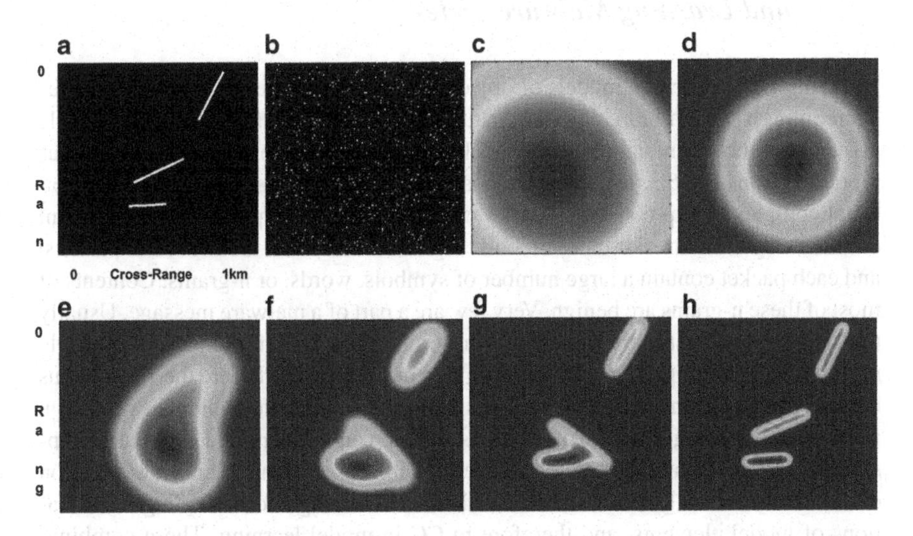

Fig. 6.1 Six scans are shown on top of each other. Three moving objects are shown in (**a**). These objects and 18 signals from these objects are below clutter and cannot be seen among 3,000 clutter points (**b**). The initial model with arbitrary parameters are shown in (**c**); because of the large variance this state is vague, corresponding to Aristotelian potentiality for objects. (**d**–**h**) show successive DL steps. The number of objects and parameters are accurately estimated in 17 steps. Standard multiple hypothesis tracking, evaluating all tracking association hypotheses, would require about 10^{2100} operations, a number too large for computation. Therefore, existing tracking systems require strong signals, with about 15 dB signal-to-clutter ratio [13]. Except for the DL tracker there is no other algorithm that can track target signals below clutter like in this example. DL successfully detected and tracked all three targets and required only 10^6 operations, achieving about 18 dB improvement (70 times) in signal-to-clutter sensitivity

objects signals are below clutter, so they are not seen. In pattern recognition usually algorithms cannot perform better than the human visual system, but DL can find the three moving objects in clutter. Beginning with a vague, uncertain state in (c), the DL process finds the objects in 17 steps. This DL process corresponds to Aristotelian process of "mind meeting matter"; the original vague state of the models corresponds to the Aristotelian potential state of forms, which in the process of interacting with measured signals turn into actualities: accurate representations of objects in figure (h). As compared to previous state of the art [13], this is an improvement of 18 dB (70 times) in sensitivity in terms of signal to clutter ratio.

This breakthrough performance of DL is achieved by avoiding logic. An essential aspect of tracking in other algorithms is an association step: making decisions about which signals are associated with which model (object or clutter). Association has been treated logically, as an essentially discrete decision. DL transformed these logical discrete decisions into continuous dynamic processes, Eqs. 6.3 and 6.4. This enabled overcoming CC and achieving this 70 times performance improvement.

6.4.1 Continuous Representation of Model Structure and Learning Malware Codes

Applying DL to learning models of objects and events in Internet network requires overcoming CC of a more complex nature in the above example. In this example associations between signals and models have been transformed into continuous representations Eq. 6.3, which makes possible the DL processes avoiding combinatorial complexity Eq. 6.4. For Internet models we face a requirement to represent continuously structures of these models. Signals in networks are moved in packets, and each packet contain a large number of symbols, words, or n-grams. Contents of most of these n-grams are benign. Very few are a part of a malware message. Usually the dangerous or destructive nature of an n-gram cannot be determined from a single n-gram. Several of them have to be assembled into a message before their dangerous content can be determined. This requires sorting through a huge number of benign n-grams and messages before a specific structure can be identified. Structural constituents of a model are considered inherently discrete elements. This view based on classical logical analysis of a model leads to considering and evaluating combinations of model elements, and therefore to CC in model learning. These combinations are of an entirely different nature than combinations of signals and models in the previous example, and the previously developed mathematical approach is not applicable here.

Below we describe a further development of DL, which turns identifying of a model structure into a continuous problem [48]. Instead of the logical consideration of a model as consisting of its elements, so that every signal or n-gram in the network either belongs to a model or does not, DL considers every n-gram as potentially belonging to a model. Starting from a vague potentiality of a model, to which every n-gram could belong, the DL learning process evolves this into a model-actuality containing definite n-grams, and not containing others.

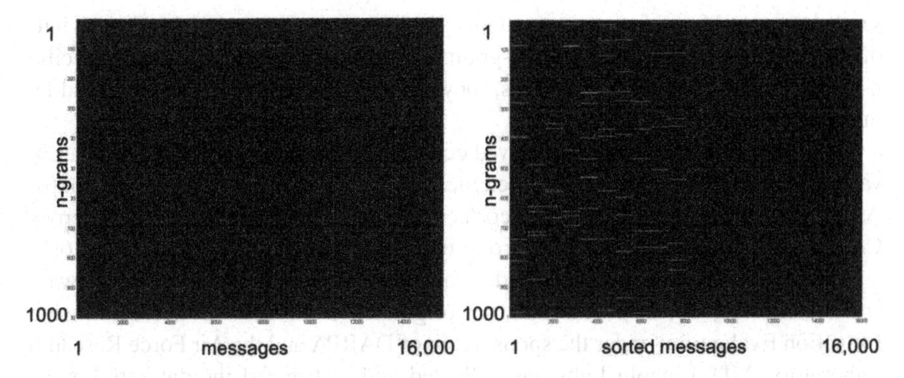

Fig. 6.2 On the left are 16,000 messages arranged in their sequential order along the horizontal axis, n. The total number of possible n-grams in the network is 1,000, they are shown along the vertical axis, j. Every message has or does not have a particular n-gram as shown by a *white* or *black dot* at the location (n,j). This figure looks like random noise corresponding to pseudo- random content of messages. On the right figure, messages are sorted so that messages having similar n-grams appear next to each other. These similar n-grams appear as white horizontal stricks. Most of message contents are pseudo-random n-grams; about a half of messages have several similar n-grams. These messages with several specific n-gram values have specific contents, they belong to certain models, and could be malware codes

We denote n-grams in the network as x(n,j), here n = 1,…N enumerates messages, and j = 1,…J enumerates n-grams. As reviously, m = 1,…M enumerates models. Model parameters, in addition to r(m) are p(m,j), potentialities of n-gram j belonging to model m. Data x(n,j) have values 0 or 1; potentialities p(m,j) start with vague value near 0.5 and in the DL process of learning they converge to 0 or 1. Mathematically this construct can be described as

$$\ell(\mathrm{n}|\mathrm{m}) = \prod_{j=1}^{J} \mathrm{p}(\mathrm{m,j})^{\mathrm{x(n,j)}} \left(1-\mathrm{p}(\mathrm{m,j})\right)^{(1-\mathrm{x(n,j)})}. \tag{6.8}$$

A model parameter p(m,j), modeling a potentiality of n-grams j being part of model m, starts the DL process with initial value near 0.5 (exact values 0.5 for all p(m,j) would be a stationary point of the DL process Eq. 6.4). Value p(m,j) near 0.5 gives potentiality values of x(n,j) with a maximum near 0.5, in other words, every n-gram has a significant chance to belong to every model. If p(m,j) converge to 0 or 1 values, these would describe which n-grams j belong to which models m.

The data used for testing this DL algorithm and the results of the analysis are shown in Fig. 6.2. We simulated 16,000 messages shown on the left. They are arranged in their sequential order along the horizontal axis, n. For this simplified example we simulated 1,000 total number of possible n-grams in the network, they are shown along the vertical axis, j. Every message has or does not have a particular n-gram as shown by a white or black dot at the location (n,j). This figure looks like random noise corresponding to pseudo- random content of messages. On the right figure, messages are sorted so that messages having similar n-grams appear next to each other. These similar n-grams appear as white horizontal strikes and reveal several groups. Most of message contents are pseudo-random n-grams; about a half of messages have several similar n-grams. These messages with several specific n-gram values have specific contents, they belong to certain models, and could be malware codes.

Since the data for this example have been simulated, we know the true number of various groups, and the identity of each message as belonging to a particular groups. All messages have been assigned correctly to its group without a single error. Convergence is very fast and took two to four iterations (or steps) to solve Eq. 6.4.

This algorithm have been applied to a publicly available data set of malware codes, KDD [7, 10, 21]. This data set originated from 1998 DARPA Intrusion Detection Evaluation; under the sponsorship of DARPA and the Air Force Research Laboratory, MIT Lincoln Labs has collected and distributed the datasets for the evaluation of computer network intrusion detection system. This data set includes 41 features extracted from Internet packets and one class attribute enumerating 21 classes of four types of attacks. Our algorithm identified all classes of malware and all malware messages without a single false alarm. (This seems to be a better performance than other published algorithms).

The machine learning based on DL achieves the fastest possible learning. This is a consequence of DL performing the maximum likelihood model estimation. This is known to lead to algorithms reaching the Cramer-Rao Bound, the information-theoretic bound on speed of learning [27, 28, 48]. Thus in the future battles of machine learning technologies between cyber-security threats and defenses, DL offers a mathematically-provable technique with the fastest adaptive capability.

6.4.2 Cognitive and Emotional Functions of DL in the Mind

In addition to being a mathematical breakthrough in several areas of Machine Learning, DL is a cognitive mathematical theory, a basis for a number of cognitive algorithms. Calling DL a cognitive mathematical theory, we mean that it mathematically models several functions of the mind-brain. These models explain functions of the mind-brain, some of which are well appreciated, others seemed mysterious. They made a number of experimentally testable predictions, some unexpected and confirmed experimentally, none have been disconfirmed. This section briefly summarizes this cognitive aspect of DL.

The first salient and unexpected prediction of dynamic logic is the process from vague to crisp as a foundation of perception and cognition. A simplified experiment confirming this prediction can be conducted by anyone in 1/2 a minute. Concentrate on an object in front of your eyes, then close the eyes and imagine the object. Imagination is usually not as clear and crisp as perception with opened eyes. It is known that imagination is produced by top-down neural signals from representations of objects stored in memory [15]. Therefore, representations are not as crisp and clear as perceptions with open eyes, representations are vague. The more abstract are imagined ideas, the vaguer are representations. This experiment have been performed using brain imaging [2, 36]. This experiment confirmed that representations are vague, and less conscious than perceptions with opened eyes. Predictions about vaguer nature of cognitive representations have been confirmed in [17].

According to the theory of instincts and emotions [12], instincts are sensor-like neural mechanisms in the mind, which measure vital bodily parameters and indicate their safe ranges to decision-making parts of the brain. The neural signals connecting instinctual and decision-making parts of the brain are emotional signals indicating satisfaction of instinctual needs. DL extended this theory toward the knowledge instinct, which measures similarity between mental models-representations and patterns in sensor signals, Eq. 6.1 [22, 27, 23]. Emotional signals measuring satisfaction of the knowledge instinct are aesthetic emotions, serving as a foundation for all human higher mental abilities, including abilities for the beautiful [26–28, 37]. Existence of this specific aesthetic emotions related to knowledge have been first postulated by Kant [14] and experimentally confirmed in [49].

DL has led to a theory of interaction between language and cognition [8, 9, 29, 32, 34, 35, 38, 39, 47, 55, 51]. This theory explains why children can talk without full understanding, why language is acquired earlier than cognition, several other mysteries of language and cognition, it predicts that language and cognition are closely connected but separate brain functions, that abstract concepts are understood mostly due to language, without full cognitive understanding. Some of these predictions have been experimentally confirmed [3, 52]. DL has led to a theory of language emotionality [35, 43] and to a theory explaining the cognitive function, origin, and evolution of musical emotions that Darwin (1871) called "the greatest mystery" [31, 33, 38, 40–42, 44, 46]. Predictions of this theory have been experimentally confirmed in [5, 18–20, 50].

6.5 Future Research

The next step of applying DL to cyber security will include learning syntactic and semantic aspects of models. Fast learning of these complicated models is necessary to counter advanced threats, including evolving malware using stealthy, mutating, self-camouflaging, and "Frankensteinian" technologies [6]. These types of threat are capable to evolve and mutate for avoiding existing anti-malware technology, operate stealthy, and assemble itself from parts of other codes (so that no "local" syntax-based detection is possible). We repeat: future malware codes will utilize machine learning technology, and countering these cyber-threats will be only possible by using a superior machine learning technology. Future cyber-security will be a battle of machine learning technologies. Here we described a step toward the machine learning technology that can be mathematically proven to reach information theoretic bounds on speed of learning [45, 48].

Cognitive and emotional aspects of DL machine learning will be used for a different type of cyber-security than the one at the focus of this chapter. It will be possible to analyze cognitive and emotional contents of network traffic, identify perpetrators and their intents, and instead of countering cyber-threats, attack the perpetrators of cyber-attacks.

References

1. Aristotle. (1995). The complete works. The revised Oxford translation, ed. J. Barnes, Princeton, NJ: Princeton Univ. Press. (Original work IV BCE).
2. Bar, M., Kassam, K.S., Ghuman, A.S., Boshyan, J., Schmid, A.M., Dale, et al. (2006). Top-down facilitation of visual recognition. USA: Proceedings of the National Academy of Sciences, 103, 449-54.
3. Binder, J.R., Westbury, C.F., McKiernan, K.A., Possing, E.T., & Medler, D.A. (2005). Distinct Brain Systems for Processing Concrete and Abstract Concepts. Journal of Cognitive Neuroscience 17(6), 1–13.
4. Blowers, M. and Williams, J. (2014). Machine Learning Applied to Cyber Operations. In Pino, R.E. (ed.) Network Science and Cybersecurity, Springer, New York, NY.
5. Cabanac, A., Perlovsky, L.I., Bonniot-Cabanac, M-C., Cabanac, M. (2013). Music and Academic Performance. Behavioural Brain Research, 256, 257-260.
6. Cisco. (2013). Annual Security Report; https://www.cisco.com/web/offer/gist_ty2_asset/Cisco_2013_ASR.pdf
7. Dua, S. & Du, X. (2011). Data Mining and Machine Learning in Cybersecurity. Taylor & Francis, Boca Raton, FL.
8. Fontanari, J.F. and Perlovsky, L.I. (2007). Evolving Compositionality in Evolutionary Language Games. IEEE Transactions on Evolutionary Computations, 11(6), pp. 758-769; doi:10.1109/TEVC.2007.892763
9. Fontanari, J.F. & Perlovsky, L.I. (2008a). How language can help discrimination in the Neural Modeling Fields framework. Neural Networks, 21(2-3), 250–256.
10. Gesher, A. (2013). Adaptive adversaries: building systems to fight fraud and cyber intruders. In Proceeding of the 19th ACM SIGKDD international conference on Knowledge discovery and data mining, Pages 1136-1136, ACM New York, NY, US.
11. Gödel, K. (2001). Collected Works, Volume I, Publications 1929–1936. Feferman, S., Dawson, J.W., Jr., Kleene, S.C., Eds.; Oxford University Press: New York, NY.
12. Grossberg, S. & Levine, D.S. (1987). Neural dynamics of attentionally modulated Pavlovian conditioning: blocking, inter-stimulus interval, and secondary reinforcement. Psychobiology, 15(3), pp.195-240.
13. Jones, J. Bradstreet, J., Kozak, M., Hughes, T., & Blount, M. (2004). Ground moving target tracking and exploitation performance measures. Pentagon Report A269234, approved for public release.
14. Kant, I. (1790/1914). Critique of Judgment, tr. J.H.Bernard, London: Macmillan & Co. Kant, 1798/1974.
15. Kosslyn, S.M. (1994). Image and Brain. Cambridge, MA: MIT Press.
16. Kovalerchuk, B., Perlovsky, L., & Wheeler, G. (2012). Modeling of Phenomena and Dynamic Logic of Phenomena. Journal of Applied Non-classical Logics, 22(1), 51-82. http://arxiv.org/abs/1012.5415
17. Kveraga, K., Boshyan, J., & M. Bar. (2007) Magnocellular projections as the trigger of top-down facilitation in recognition. Journal of Neuroscience, 27, 13232-13240.
18. Masataka, N. & Perlovsky, L.I. (2012a). Music can reduce cognitive dissonance. Nature Precedings: hdl:10101/npre.2012.7080.1; http://precedings.nature.com/documents/7080/version/1
19. Masataka, N. & Perlovsky, L.I. (2012b). The efficacy of musical emotions provoked by Mozart's music for the reconciliation of cognitive dissonance. Scientific Reports 2, Article number: 694 doi:10.1038/srep00694 http://www.nature.com/srep/2013/130619/srep02028/full/srep02028.html
20. Masataka, N. & Perlovsky, L.I. (2013). Cognitive interference can be mitigated by consonant music and facilitated by dissonant music. Scientific Reports 3, Article number: 2028 (2013) doi:10.1038/srep02028; http://www.nature.com/srep/2013/130619/srep02028/full/srep02028.html

21. Mugan, J. (2013). A developmental approach to learning causal models for cyber security. Proc. SPIE 8751, Machine Intelligence and Bio-inspired Computation: Theory and Applications VII, 87510A (May 28, 2013); doi:10.1117/12.2014418
22. Perlovsky, L.I. (1987). Multiple sensor fusion and neural networks. DARPA Neural Network Study, 1987.
23. Perlovsky, L.I. & McManus, M.M. (1991). Maximum Likelihood Neural Networks for Sensor Fusion and Adaptive Classification. Neural Networks 4 (1), 89-102.
24. Perlovsky, L.I. (1996). Gödel Theorem and Semiotics. Proceedings of the Conference on Intelligent Systems and Semiotics'96. Gaithersburg, MD, v.2, pp. 14-18.
25. Perlovsky, L.I. (1998). Conundrum of Combinatorial Complexity. IEEE Trans. PAMI, 20(6) pp. 666-670.
26. Perlovsky, L.I. (2000). Beauty and Mathematical Intellect. Zvezda, 2000(9), 190-201 (Russian)
27. Perlovsky, L.I. (2001a). Neural Networks and Intellect: using model-based concepts. Oxford University Press, New York, NY (3rd printing).
28. Perlovsky, L. I. (2001b). Mystery of sublime and mathematics of intelligence. Zvezda, 2001(8), 174-190, St. Petersburg.
29. Perlovsky, L.I. (2004). Integrating Language and Cognition. IEEE Connections, Feature Article, 2(2), 8-12.
30. Perlovsky, L.I. (2006a). Toward Physics of the Mind: Concepts, Emotions, Consciousness, and Symbols. Phys. Life Rev. 3(1), 22-55.
31. Perlovsky, L.I. (2006c). Music – The First Principle. Musical Theatre, http://www.ceo.spb.ru/libretto/kon_lan/ogl.shtml
32. Perlovsky, L.I. (2007). Evolution of Languages, Consciousness, and Cultures. IEEE Computational Intelligence Magazine, 2(3), 25-39
33. Perlovsky, L.I. (2008). Music and Consciousness, Leonardo, Journal of Arts, Sciences and Technology, 41(4), pp.420-421.
34. Perlovsky, L.I. (2009a). Language and Cognition. Neural Networks, 22(3), 247-257. doi:10.1016/j.neunet.2009.03.007.
35. Perlovsky, L.I. (2009b). Language and Emotions: Emotional Sapir-Whorf Hypothesis. Neural Networks, 22(5-6); 518-526. doi:10.1016/j.neunet.2009.06.034
36. Perlovsky, L.I. (2009c). 'Vague-to-Crisp' Neural Mechanism of Perception. IEEE Trans. Neural Networks, 20(8), 1363-1367.
37. Perlovsky, L.I. (2010a). Intersections of Mathematical, Cognitive, and Aesthetic Theories of Mind, Psychology of Aesthetics, Creativity, and the Arts, 4(1), 11-17. doi: 10.1037/a0018147.
38. Perlovsky L.I. (2010b). Physics of The Mind: Concepts, Emotions, Language, Cognition, Consciousness, Beauty, Music, and Symbolic Culture. WebmedCentral PSYCHOLOGY 2010;1(12):WMC001374; http://arxiv.org/abs/1012.3803
39. Perlovsky, L.I. (2010c). Joint Acquisition of Language and Cognition; WebmedCentral BRAIN;1(10):WMC00994; http://www.webmedcentral.com/article_view/994
40. Perlovsky, L.I. (2010d). Musical emotions: Functions, origin, evolution. Physics of Life Reviews, 7(1), 2-27. doi:10.1016/j.plrev.2009.11.001
41. Perlovsky, L.I. (2012a). Cognitive function, origin, and evolution of musical emotions. Musicae Scientiae, 16(2), 185 – 199; doi: 10.1177/1029864912448327.
42. Perlovsky, L.I. (2012b). Cognitive Function of Music, Part I. Interdisciplinary Science Reviews, 37(2), 129–42.
43. Perlovsky, L.I. (2012c). Emotionality of Languages Affects Evolution of Cultures. Review of Psychology Frontier, 1(3), 1-13. http://www.academicpub.org/rpf/paperInfo.aspx?ID=31
44. Perlovsky, L.I. (2013a). A challenge to human evolution – cognitive dissonance. Front. Psychol. 4:179. doi: 10.3389/fpsyg.2013.00179; http://www.frontiersin.org/cognitive_science/10.3389/fpsyg.2013.00179/full
45. Perlovsky, L.I. (2013b). Learning in brain and machine - complexity, Gödel, Aristotle. Frontiers in Neurorobotics; doi: 10.3389/fnbot.2013.00023; http://www.frontiersin.org/Neurorobotics/10.3389/fnbot.2013.00023/full

46. Perlovsky, L.I. (2013c). Cognitive Function of Music, Part II. Interdisciplinary Science Reviews, 38(2), 149-173.
47. Perlovsky, L.I. (2013d). Language and cognition – joint acquisition, dual hierarchy, and emotional prosody. Frontiers in Behavioral Neuroscience, 7:123; doi:10.3389/fnbeh.2013.00123; http://www.frontiersin.org/Behavioral_Neuroscience/10.3389/fnbeh.2013.00123/full
48. Perlovsky, L.I., Deming R.W., & Ilin, R. (2011). Emotional Cognitive Neural Algorithms with Engineering Applications. Dynamic Logic: from vague to crisp. Springer, Heidelberg, Germany.
49. Perlovsky, L. I., Bonniot-Cabanac, M.-C., Cabanac, M. (2010). Curiosity and Pleasure. WebmedCentral PSYCHOLOGY 2010;1(12):WMC001275; http://www.webmedcentral.com/article_view/1275; http://arxiv.org/ftp/arxiv/papers/1010/1010.3009.pdf
50. Perlovsky, L.I., Cabanac, A., Bonniot-Cabanac, M-C., & Cabanac, M. (2013). Mozart Effect, Cognitive Dissonance, and the Pleasure of Music. ArXiv 1209.4017; Behavioural Brain Research, 244, 9-14.
51. Perlovsky, L.I. & Ilin, R. (2010). Neurally and Mathematically Motivated Architecture for Language and Thought. Special Issue "Brain and Language Architectures: Where We are Now?" The Open Neuroimaging Journal, 4, 70-80. http://www.bentham.org/open/tonij/openaccess2.htm
52. Price, C.J. (2012). A review and synthesis of the first 20 years of PET and fMRI studies of heard speech, spoken language and reading. NeuroImage, 62, 816–847.
53. Shabtai, A., Moskovitch, R., Feher, C., Dolev, S., & Elovici, Y. (2012). Detecting unknown malicious code by applying classification techniques on OpCode patterns, Security Informatics, 1:1; http://www.security-informatics.com/content/1/1/1.
54. Singer, R.A., Sea, R.G. & Housewright, R.B. (1974). Derivation and Evaluation of Improved Tracking Filters for Use in Dense Multitarget Environments, IEEE Transactions on Information Theory, IT-20, 423-432.
55. Tikhanoff. V., Fontanari, J. F., Cangelosi, A. & Perlovsky, L. I. (2006). Language and cognition integration through modeling field theory: category formation for symbol grounding. In Book Series in Computer Science, v. 4131, Heidelberg: Springer.
56. Vityaev, E.E., Perlovsky, L.I., Kovalerchuk, B.Y., Speransky, S.O. (2011). Probabilistic dynamic logic of the mind and cognition, Neuroinformatics, 5(1), 1-20.
57. Vityaev, E.E., Perlovsky, L.I., Kovalerchuk, B. Y., & Speransky, S.O. (2013). Probabilistic dynamic logic of cognition. Invited Article. Biologically Inspired Cognitive Architectures 6, 159-168.

Chapter 7
Towards Neural Network Based Malware Detection on Android Mobile Devices

Wei Yu, Linqiang Ge, Guobin Xu, and Xinwen Fu

7.1 Introduction

The rapid growth of smart mobile devices has led to a renaissance for mobile services. These devices can augment cognitive abilities with multi-function applications related to web, education, travel, game, financial, and many others. For example, face recognition applications can help identify or verify a person to enhance human cognitive abilities. The Android platform is an open source operating system for smart mobiles and provides services, including security configuration, process management, and others [1]. With 48 % of smartphone subscribers using Android mobiles, Android leads the smartphone market in the U.S. [2].

Nonetheless, the popularity of Android mobile devices has led to enormous security challenges. Malware, as a malicious application that can be installed on mobile devices, can gain access to these devices and collect user sensitive information. Malware has proven to be a serious problem for the Android platform because malicious applications can be distributed to mobile devices through an application market. From the defender's perspective, how to effectively detect malware and enhance the cognitive performance of users and system administrators becomes a challenging issue. Traditional static analysis techniques heavily rely on capturing malicious characteristics and bad code segments embedded in software. This makes it infeasible to deal with a large population of unknown malware. Hence, it is critical to develop a machine learning-based system that can dynamically learn the

W. Yu (✉) • L. Ge • G. Xu
Department of Computer & Information Sciences, Towson University, Towson, MD, USA
e-mail: wyu@towson.edu; lge2@students.towson.edu; gxu2@students.towson.edu

X. Fu
Department of Computer Science, University of Massachusetts, Lowell, MA, USA
e-mail: xinwenfu@cs.uml.edu

R.E. Pino et al. (eds.), *Cybersecurity Systems for Human Cognition Augmentation*, Advances in Information Security 61,
DOI 10.1007/978-3-319-10374-7_7

behavior of malware and augment the human cognition process of defending against malware attacks in the battle of mobile security.

In this chapter, we propose an Artificial Neural Network (ANN)-based malware detection system that uses both permissions and system calls to detect unknown malware. In our system, we consider two types of ANNs: Feedforward Neural Networks (FNN) to learn the patterns of permissions and Recurrent Neural Networks (RNN) to understand the structure of system calls. Permission requests are collected from applications to distinguish between benign applications and malware. We also collected system calls associated with application execution to capture the runtime behaviors of benign applications and malware. Through the training process, the ANN can learn the anomaly behaviors of malware in terms of permission requests and system calls. The resulting model can be further used to detect unknown malware. To evaluate the effectiveness of our malware detection system, we used real-world malware and benign applications to conduct experiments on Android mobile devices. The resulting data shows that our system can effectively detect malware.

The remainder of the chapter is organized as follows: We introduce the ANN in Sect. 7.2 and two types of data sources for malware detection: permission and system calls in Sect. 7.3. In Sect. 7.4, we present our ANN-based malware detection system. Experimental results were demonstrated to validate the effectiveness of our proposed detection system in Sect. 7.5. We then discuss the issues related to our work in Sect. 7.6. We review related work in Sect. 7.7 and conclude the chapter in Sect. 7.8.

7.2 Artificial Neural Networks

We consider ANN to conduct malware detection. Generally speaking, a neural network refers to a network or circuit that mimics the structure and behavior of biological neurons [3]. The parameters of a neural network are set through a training process that uses known data sets as inputs. After that, the trained neural network can be used as a classifier to conduct detection.

7.2.1 Feedforward Neural Networks (FNN)

FNNs are a well-known and widely used type of neural network [4–8]. An FNN consists of a certain number of layers and a number of units called artificial neurons or nodes that are organized in layers. In a typical setting, an FNN has an input layer, an output layer, and one or more hidden layers between the input and the output layer. In an FNN, all data and computation flows are in one direction: from input to output data. Except for input units, each unit in a layer is connected to all the units in the previous layer and receives inputs directly from the nodes in the previous layer. Each connection may have a different strength or weight. During the training process, the

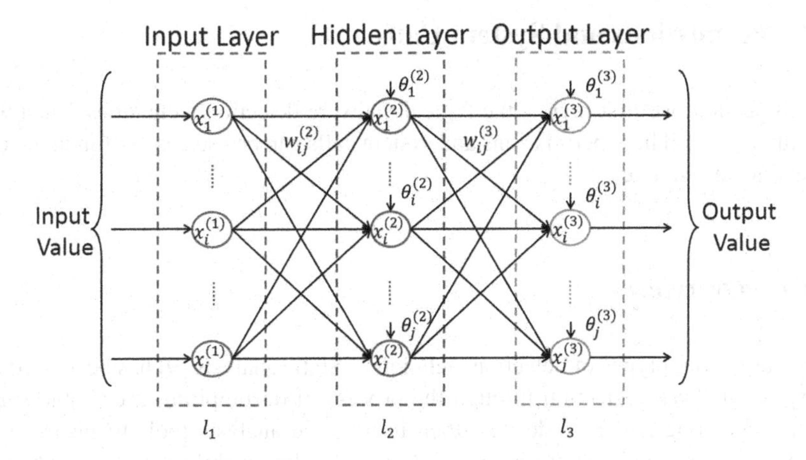

Fig. 7.1 A typical structure of FNN

weight can be adjusted through learning algorithms such as BackPropagation (BP). The typical structure of an FNN is illustrated in Fig. 7.1.

Here, l represents the layer of the FNN, where $l=1$ is for the input layer, $l=2$ is for the hidden layer, and $l=3$ is for the output layer. In principle, the output values are compared with the correct answer to compute the value of a predefined error-function that is then sent back through the network. With the backward propagation errors between real and estimated values from the output layer to the hidden layer and from the hidden layer to the input layer, errors in each layer can be estimated and the assigned weights $\omega_{ij}{}^{(l)}$ can be updated correspondingly. After repeating this procedure many times, the neural network eventually reaches a state where the computed error is small. At this moment, the training process is complete.

7.2.2 Recurrent Neural Networks (RNN)

Unlike the FNN, the fundamental feature of an RNN is that the network contains at least one feedback connection. This makes an RNN useful for handling temporal classification problems or learning sequences. Similar to an FNN, an RNN consists of a number of units and multiple layers: input layer, output layer, and one or more hidden layers. When the data is fed to an RNN, a state activation is generated in the hidden layers. In the next time slot, the previous state activation is fed back to the hidden layer, combining with new input data. During the training process, the weight of unit connections and feedback connections can be adjusted through learning algorithms such as Back Propagation Through Time (BPTT). The BP algorithm used in an FNN cannot be directly applied to an RNN because of the inherent cycles present. Hence, BPTT unfolds the network over time, eliminating cycles and allowing the neural network to be trained as if it consists of several connected FNNs where the BP algorithm can be used.

7.3 Permissions and System Calls

In this section, we first review the typical malware detection techniques. Then we examine in detail how permissions and system calls can be used as the fundamental detection data source.

7.3.1 Overview

There are several types of detection techniques. Static analysis [9] has been used to carry out malware detection through the process of decompiling executable software, generating source code, and then using code analysis tools to inspect the recovered source code. Static analysis is limited by the capability of code analyzers and can only deal with applications that involve a small number of permissions and system calls.

Permission and dynamic analysis schemes are promising techniques to defend against a large class of unknown malware. To be specific, permission-based detection sets security policy rules. When an application is installed, the permission-based detection extracts security configurations and checks them against security policy rules [10]. Conversely, dynamic analysis-based detection [11] executes the mobile application and monitors the applications dynamic behavior. Based on the runtime behavior, the malware can be detected. As malicious behavior is always difficult to hide and can be used as a feature to identify malware, we can use ANN techniques to accurately characterize the behavior of applications.

7.3.2 Permissions

Android provides third-party applications that have the capability of accessing resources such as phone hardware, settings, user data, and others through permissions. For example, the INTERNET permission allows applications to open network connections. Each application must declare in advance what permissions it requires, and users are notified during the installation about the permissions that it will obtain. Users can cancel the installation process if they do not want to grant a permission to the application, but they might not have the knowledge to determine which permissions should be requested by and granted by a particular application. Usually, different types of applications request reasonable permissions. Nonetheless, even an application requesting a reasonable permission might conduct malicious behavior. For example, a social network application that requests to only access the contact may additionally copy contacts personal information to a remote server.

To show the potential of using permissions to detect malware, we investigated the distribution of permissions requested by electronic books - one class of

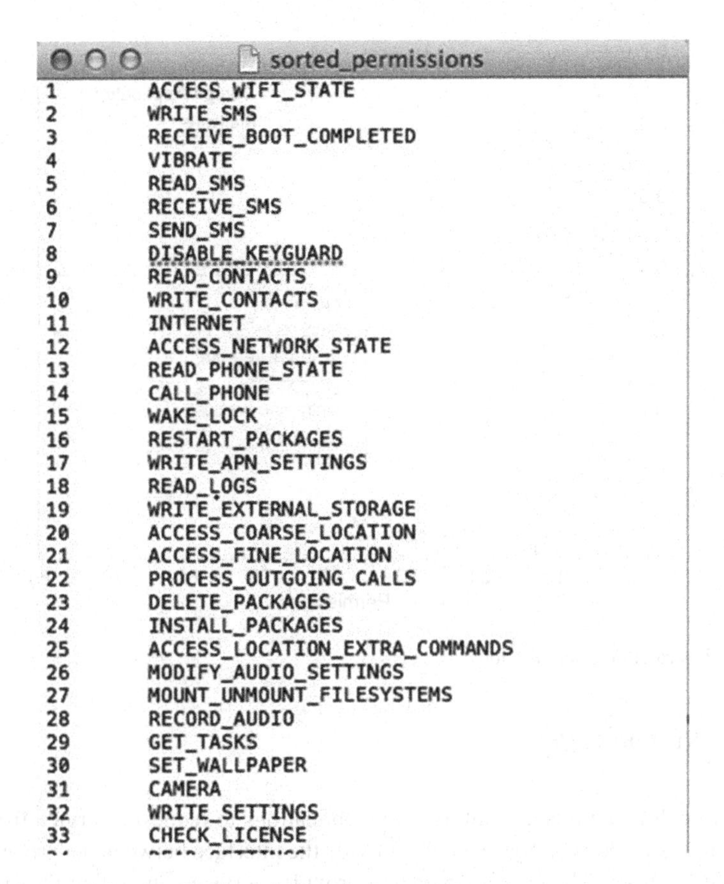

Fig. 7.2 An example of mapped permissions

applications. We installed 96 benign applications from Google Play and used 92 digital book malware samples from the Android Malware Genome Project (http://www.malgenomeproject.org/). For each Android application, we extracted permissions from the corresponding application package (APK) file. The details of the retrieving process will be presented in Sect. 7.4.1. We define each captured permission as one feature and map it to an integer. Figure 7.2 shows an example of mapped permissions.

After retrieving the permissions from all applications, the distribution of permissions can be computed. One such example is shown in Fig. 7.3. As we can see, most malware samples heavily request permissions 1–20, which are WRITE SMS, SEND SMS, READ CONTACT, etc. We can conclude that electronic book applications that request permissions 1–20 are probably malware. Hence, the permissions requested by an application can be used to recognize whether the application contain malware.

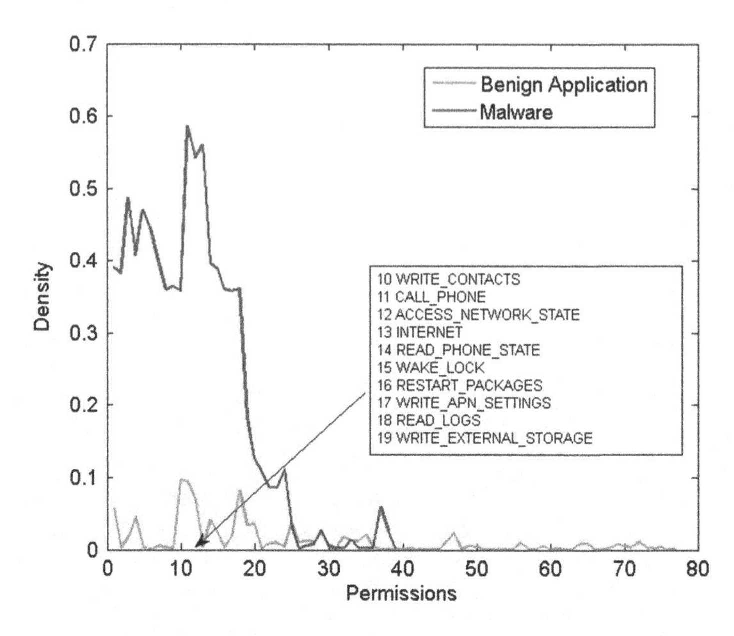

Fig. 7.3 Distribution of permissions

7.3.3 *System Calls*

A system call is the mechanism used by applications to request a service from the operating system kernel. System calls provide the interface between the process and operating systems. The operating system provides services, including the creation and execution of new processes and access control of resources. The sequence of system calls occurs consecutively over time and can capture actions performed by applications during execution. As system calls provide an essential interface between the application and operating system, we shall examine system calls to capture the runtime behavior of the interactions between applications and the operating system.

7.4 An ANN-Based Malware Detection System

We now present the workflow of our proposed ANN-based malware detection system as shown in Fig. 7.4. We would like to emphasize that the workflow is general and can be used for both permission-based detection and system call- based detection. In the offline training phase, we first collected real-world benign and malicious applications. Next, we executed the collected applications and dumped the data sources. In order for machine learning algorithms to learn the feature patterns of

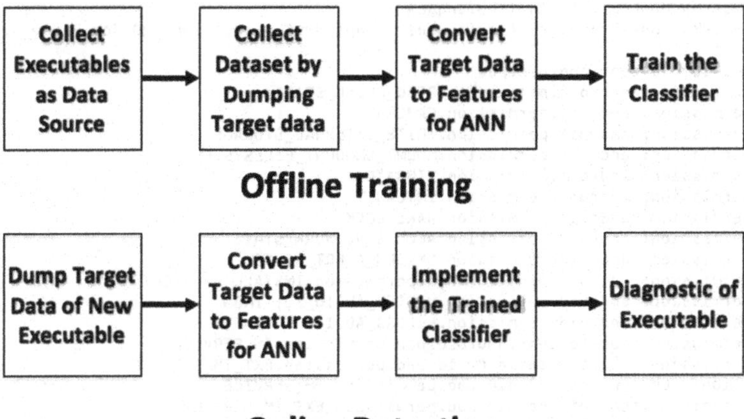

Fig. 7.4 Workflow

malware and benign applications, all data sources needed to be parsed and mapped to the format required by the FNN and RNN algorithms described in Sect. 7.2. Using the mapped data as input, we then trained the neural network. In the online detection phase, we dumped the data sources from new applications and the trained neural network would be used to determine whether the new application is malware or benign. As permissions and system calls contain different features and have different formats, we first introduce permission-based detection and then system call-based detection in the following subsections.

7.4.1 Permission-Based Detection

Offline Training We now discuss the steps used for the offline training process.

Step 1: **Data source collection and classification**. The first step in the offline training phase is to collect the data source from the executing applications. With real-world benign applications and malware samples, we consider that applications in the same category should exhibit similar activities and we use such activities to learn the anomaly profile. Based on these learned profiles, we can categorize applications as benign or malicious.

Step 2: **Dumping Permissions of Data source**. Using the benign application and malware samples, we dump the permissions requested by each application. In the Android system, all permissions are included in the *Android-Manifest.xml* file. After collecting application *apk* files, we use a known reverse engineering tool *Android Asset Packaging Tool* (*aapt*) to reconstruct the source code and obtain the AndroidManifest.xml file for each application. An example is shown below:

```
Linqiangs-MacBook-Pro:tools linqiangge$
Linqiangs-MacBook-Pro:tools linqiangge$ ./aapt dump permissions QQ_for_Pad_v_1.9.3.a
pk
package: com.tencent.android.pad
uses-permission: android.permission.READ_PHONE_STATE
uses-permission: android.permission.CAMERA
uses-permission: android.permission.WRITE_EXTERNAL_STORAGE
uses-permission: android.permission.MOUNT_UNMOUNT_FILESYSTEMS
uses-permission: android.permission.VIBRATE
uses-permission: android.permission.INTERNET
uses-permission: android.permission.WAKE_LOCK
uses-permission: android.permission.ACCESS_NETWORK_STATE
uses-permission: android.permission.SYSTEM_ALERT_WINDOW
uses-permission: com.android.launcher.permission.INSTALL_SHORTCUT
uses-permission: android.permission.MODIFY_AUDIO_SETTINGS
uses-permission: android.permission.RECORD_AUDIO
uses-permission: com.tencent.android.pad.permission.IM_SERVICE
uses-permission: com.tencent.android.pad.permission.EXT_IM_SERVICE
permission: com.tencent.android.pad.permission.IM_SERVICE
permission: com.tencent.android.pad.permission.EXT_IM_SERVICE
permission: com.tencent.android.pad.permission.WRITE_SETTINGS
permission: com.tencent.android.pad.permission.READ_SETTINGS
Linqiangs-MacBook-Pro:tools linqiangge$ []
```

Fig. 7.5 Dumping permissions

```
<manifest xmlns:
android="http://schemas.android.com/apk/res/android"
package="com.android.app.QQ_for_Pad_v_1.9.3" >
A: android:versionCode(0x0101021b)=(type 0x10)0x7
A: android:versionName(0x0101021c)="2.1-update1"
A: package="com.android.spare_parts"
<uses-permission android:name="android.permission.READ_PHONE_
STATE"/> <uses-permission android:name="android.permission.CAMERA"/>
… </manifest>
```

We then use the command *aapt* dump permission to collect all permissions requested by each application. Figure 7.5 shows an example of the dumping process and the corresponding result.

Step 3: **Feature extraction**. Next, we collect a set of files where each file consists of permissions requested by one application. For training, we process the data and map them to the format required by the ANN. To this end, we developed a mapping algorithm to convert the original permissions into usable input. As described previously, we use Algorithm 7.1 to define each permission as one feature and assign an integer to each feature.

Using the example shown in Fig. 7.5, we now explain Algorithm 7.1. In this algorithm, we care about the feature (i.e., permission name) and the feature value, defined as whether it was requested by the application. Note that one permission can be requested only once by an application. If a particular permission is requested, its feature value is 1; otherwise its feature value is 0. After the first *for loop* of Algorithm 7.1, we obtain the output shown in Fig. 7.6.

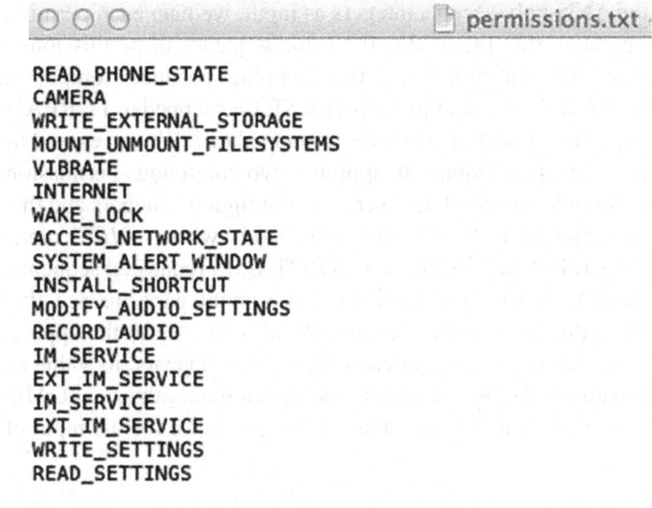

permissions.txt

```
READ_PHONE_STATE
CAMERA
WRITE_EXTERNAL_STORAGE
MOUNT_UNMOUNT_FILESYSTEMS
VIBRATE
INTERNET
WAKE_LOCK
ACCESS_NETWORK_STATE
SYSTEM_ALERT_WINDOW
INSTALL_SHORTCUT
MODIFY_AUDIO_SETTINGS
RECORD_AUDIO
IM_SERVICE
EXT_IM_SERVICE
IM_SERVICE
EXT_IM_SERVICE
WRITE_SETTINGS
READ_SETTINGS
```

Fig. 7.6 An example of permissions

Input: Original Android Permissions stored in raw data
Output: Permission Feature Vector **A**
n = gram number; **A**=[]; t=Total number of permissions
foreach *file in raw data folder* do
  foreach *line in file* do
    remove all information except the permissions name;
  end
  store file in permissions name folder;
end
foreach *file in permissions name folder* do
  foreach *line in file* do
    map permission name to integers i as feature name;
  end
  store file in mapped-integer data folder;
end
foreach *m=1:t* do
  A[m]= 0;
end
foreach *i in mapped-integer data folder* do
  A[i]=1;
end
if *n>1* ***then***
  map to n-gram format;
end

Algorithm 7.1 Permission Mapping Algorithm

Because the ANN only accepts integers as input, we map each permission name to an integer number after processing the name sequence of permissions. After the second *for loop* in Algorithm 7.1, the mapping produces output similar to "*01,02,03,06,09,15,20*". As examples, INTERNET is mapped to 11, READ PHONE STATE is mapped to 13, and SEND SMS is mapped to 7. We can extend this idea to use 2-gram as a detection feature by applying two contiguous permissions instead of one. As an example, we combine every two contiguous integers and the mapping produces output similar to "*0102,0203,0304,0405*" where "*0102*" represents the permissions ACCESS WIFI STATE and WRITE SMS requested sequentially.

After we have the input to the ANN mapped as an integer sequence, the next step is to obtain the value for the each feature. Recall that we use the appearance of a permission as the feature value. For each feature that appears, its value is assigned as 1. For the features that do not appear, we assign their values as 0. After the last two *for loops* in Algorithm 7.1, we obtain a feature vector for the input of ANN as follows:

```
  1,0,0,0,0,0,0,0,0,0,1,0,1,0,1,0,0,0,1,0,0,0,1,0,1,0, 0,0,0,0,1,
0,0,0,0,0,0,0,0,0,0,0,0,0,0,0,0,0,0,0,0,0, 0,0,0,0,0,0,0,0,0,0,0,
0,0,0,0,0,0,0,0,0,0,0,0,1,0
```

Step 4: **Classifier learning**. In this step, we use the learning module established in the neural network to learn the application behavior from training data. We input the feature vectors to the Matlab Neural Network Toolbox built-in Matlab R2013a (8.1.0.604) to implement permission-based detection. We set the number of nodes in the hidden layer to 10 and then 20.

Online Detection The workflow of the online detection phase is similar to the one described in the offline training phase. Similarly, to classify an application, the first step is to dump permissions and map the permission sequence to the format required by the ANN. We can then use the trained ANN to determine whether a new application is either malware or benign. We use the established ANN and test data as input from new applications. The test file has the same format as the training file, which consists of the feature vector associated with each application. The online detection process outputs the result file which contains the classification result. In our implementation, the result is either +1 or −1. Here, when the number is positive, the ANN classifies it as a benign application; when the number is negative, the ANN classifies it as malware.

7.4.2 System Call-Based Detection

The workflow of the detection system based on system calls is similar to the detection system based on permissions. The major difference is to use a different data source. In the following, we briefly introduce the workflow of system call- based detection.

Offline Training As before, we now discuss the steps for offline training.

Step 1: **Data set collection and classification**. The first step is to collect the data set. After we collect real-world benign applications and malware samples, we categorize them into different groups.

Step 2: **System calls recording**. We record the system calls used by our benign applications and malware samples by applying a known tool *Strace*. In order to install *Strace*, we use the *Nexus Root Tookit v1.6.2* to obtain root permission on Android devices. Next, we run *Strace* and capture the system calls used by the benign applications and malware. To install malware on an Android device from a remote computer, we use the *Android Debug Bridge* (*ADB*).

Step 3: **Feature extraction**. We then record a set of files where each file contains the system calls generated by each executed application. To use the ANN, we need to process the data and map them to the required format described previously. Using Algorithm 7.1, we map each system call named to an integer. As an example, *clock-gettime* is mapped to 1, *recvfrom* is mapped to 5, and *ioctl* is mapped to 7. Again, we can extend this idea to use 2-gram as a detection feature by applying two contiguous system calls as a detection feature instead of one. To construct the mapping for 2-gram, we combine each pair of contiguous integers and generate output similar to "*0101 0101 0105 0507 0701 0117 1717 1717 1717 1706*" where "*0105*" represents system calls *clock-gettime* and *recvfrom* executed sequentially. We then capture the density of system calls by computing the ratio of the number of instances of each system call to the total number of system calls generated by the application. We can then express a feature and its value as feature: value such as "*1:0.2283 2:0.0369 3:0.0387 4:0.0267 5:0.0848*" where feature 1 has a density of 0.2283, feature 2 has a density of 0.0369, etc.

Step 4: **Classifier learning**. This step is the same as Step 4 for permission-based detection. Afterwards, we have completed the training process of the ANN and are ready to use it to conduct online detection.

Online Detection The workflow of the online detection phase is similar to the one in the offline training phase. Similarly, to classify an application, we execute it, dump the system calls, and map the sequence of system calls to the format required by the ANN. Using the ANN established through the offline training phase, we can determine whether a new application is malware or benign.

7.5 Performance Evaluations

Using real-world malware and benign applications collected on the Android platform, we show the effectiveness of our developed detection system. We installed 96 benign software applications from Google Play and evaluated 92 digital book malware samples from the *Android Malware Genome Project* (http: //www.malgenomeproject.org/).

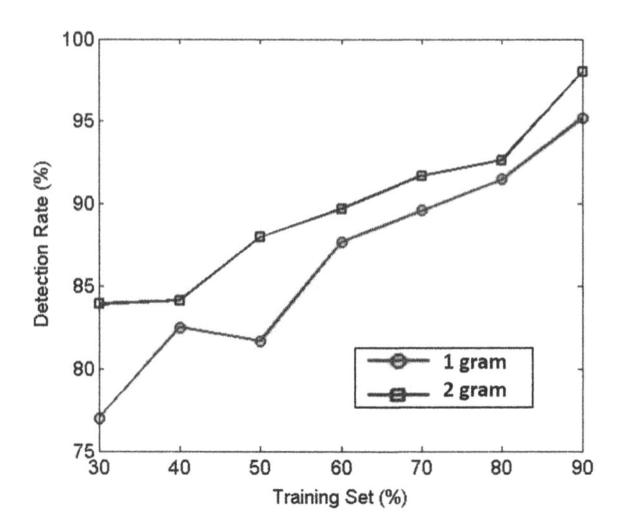

Fig. 7.7 Detection rate for permission based detection vs. Training set ratio (FNN with 10 nodes)

We installed and executed applications on the *Sumsang Galaxy Nexus* and *Google Nexus 7* smartphones in our experiments. First, we collected and transmitted each application's permission requests and system calls to a remote computer which conducted both the offline and online detection processes described in Sect. 7.4. A *Samsung Notebook NP700G* equipped with Intel Core i7 2.40GHZ processor, 16GB RAM, and 320GB hard drive served as our detection computer. Again, we used the Matlab Neural Network Toolbox built-in Matlab R2013a (8.1.0.604) that contains both of the FNN and RNN implementations used in our experiments. The number of hidden nodes in the FNN and the RNN are set to 10 and then 20.

With a larger training set, more information can be used to train the ANN classifier, leading to higher detection accuracy. To validate this hypothesis, we let $p \in [0, 1]$ which define the training set ratio as the ratio of the number of training samples to the total number of samples. If n is the number of total applications then np is the number of applications used for training and $n(1-p)$ is the number of applications used to validate the accuracy of the trained ANN. To measure the effectiveness of our detection system, we define the detection rate as the probability of correctly classifying the malware. That is, the ratio of the number of malware correctly detected to the total number of malware samples. We also define the error rate as the probability of falsely classifying applications. That is, the ratio of the number of applications falsely classified to the total number of applications.

Permission-Based Detection: Figure 7.7 illustrates the relationship between the detection rate and the training set ratio in terms of the length of grams when an FNN with ten hidden nodes is used. As we can see, in general, the detection rate rises as the training set ratio increases. The permission-based detection with 2-gram data as input can achieve a better detection rate than the permission-based detection with 1-gram data as input. For example, when the training set ratio is 60 %, the detection rate reaches almost 90 % when 2-gram are used while the detection rate is 85 %

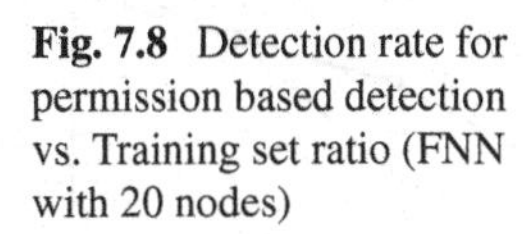

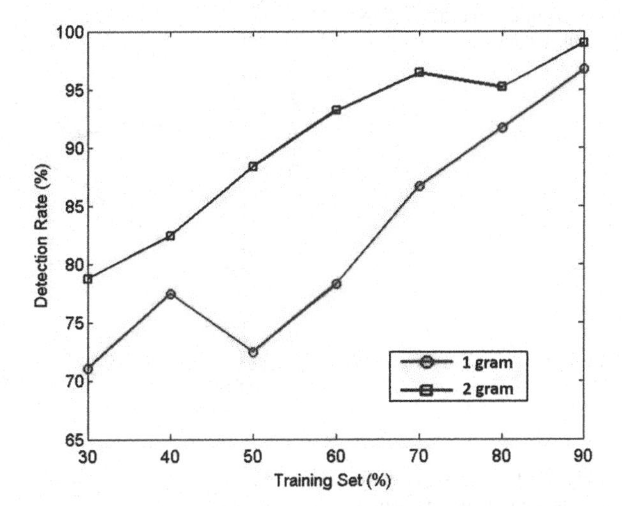

Fig. 7.8 Detection rate for permission based detection vs. Training set ratio (FNN with 20 nodes)

when 1-gram are used. As we expected, when using more training data, more knowledge of malware can be obtained, leading to increased detection accuracy.

Figure 7.8 shows the detection rate versus training set ratio when the number of hidden nodes of the FNN is set to 20. Similar to Fig. 7.7, as we increase the size of the training set, the detection rate increases. Like before, detection using 2-gram data as input achieves better performance than detection using 1-gram data as input. In the case of 2-gram data as input, when the training set ratio is higher than 50 %, the FNN with 20 hidden nodes performs better than the one with 10 hidden nodes. We also observed that, in the case of 1-gram data as input, the FNN with 10 hidden nodes performs better than the FNN with 20 hidden nodes. One reason may be caused by limited malware samples.

Figure 7.9 illustrates the result of an RNN with 10 hidden nodes. In comparison with Fig. 7.7, we can see that the FNN achieves better performance in both the 1-g and 2-gram cases than when using the RNN. Hence, we conclude that the FNN is more effective for permissions-based detection.

System Call-Based Detection: Figures 7.10, 7.11 and 7.12 illustrate the relationship between the detection rate and training set ratio in terms of the length of data grams when we take system calls as input. Similar to the permission-based detection shown in Figs. 7.7, 7.8 and 7.9, when more samples are used in the training process, a higher detection rate can be achieved. For example, when we use a training set of 90 %, both the FNN and the RNN achieved detection rates of more than 93 %. When the hidden nodes are set to 10, the RNN obtains better detection accuracy than the FNN for both permission-based and system call-based detection.

We also study the accuracy of our detection system using another metric: error rate. We expect that with a larger training set, our detection will produce a lower error rate. Figures 7.13 and 7.14 illustrate the relationship between error rate and the training set ratios when we take permissions and system calls as inputs to an FNN and an RNN. In our evaluation, we selected two scenarios to validate that our detection system obtains low error rates; other scenarios are essentially similar.

Fig. 7.9 Detection rate for permission based detection vs. Training set ratio (RNN with 10 nodes)

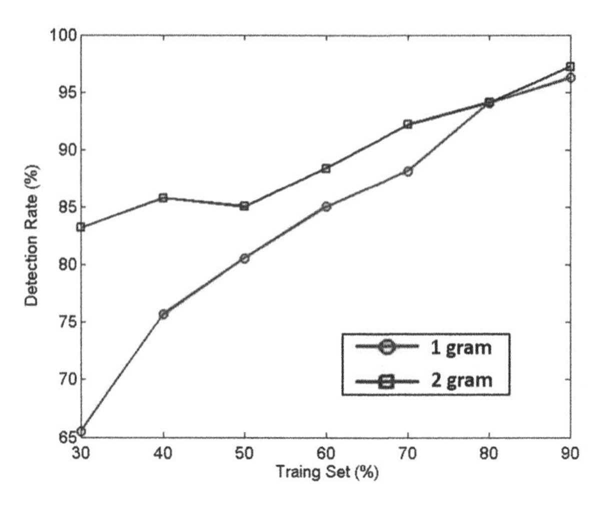

Fig. 7.10 Detection rate for system call based detection vs. Training set ratio (FNN with 10 nodes)

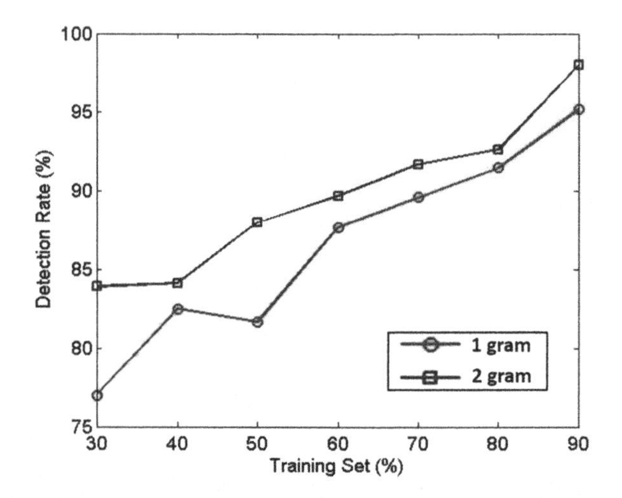

Fig. 7.11 Detection rate for system call based detection vs. Training set ratio (FNN with 20 nodes)

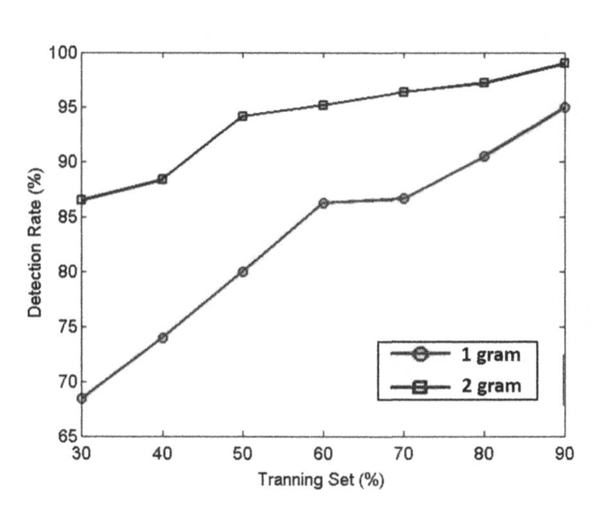

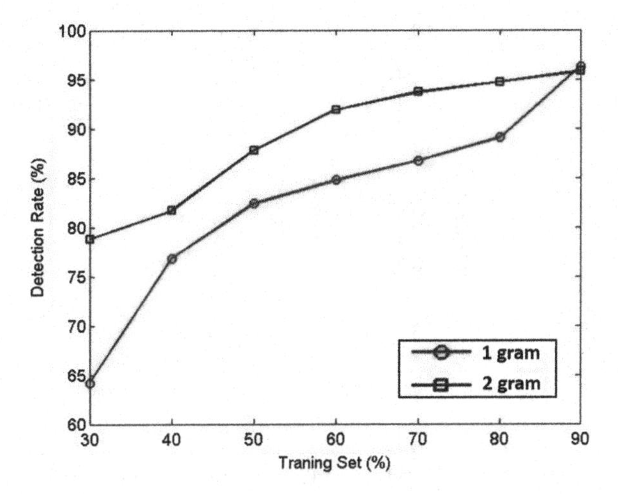

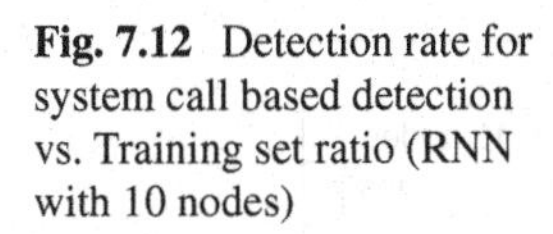

Fig. 7.12 Detection rate for system call based detection vs. Training set ratio (RNN with 10 nodes)

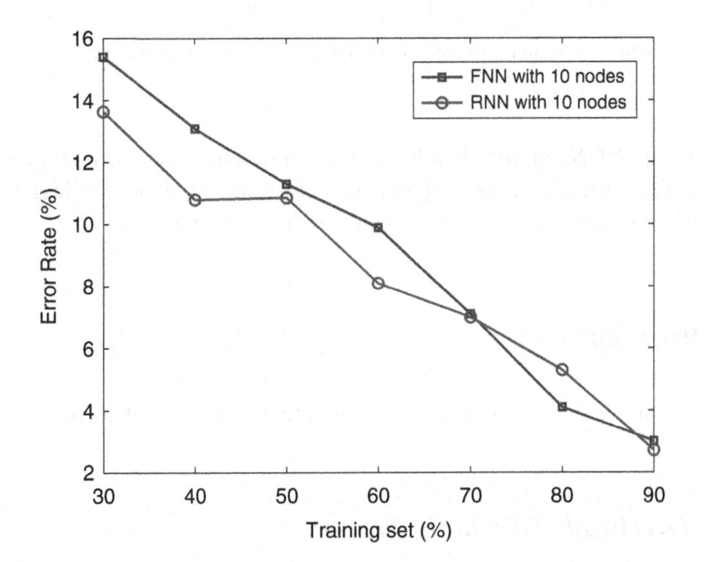

Fig. 7.13 Error rate for permission based detection vs. Training set ratio (1-g)

We used 1-gram for data input and set the hidden layer of the FNN and RNN to contain ten nodes. We have several observations from Figs. 7.13 and 7.14. First, for both permissions-based and system call-based detection, the error rates of both the FNN and RNN decrease as the training set ratio increases. This can be explained by observing that as we use more data in the training process, the FNN and RNN have a better chance to learn input data. This leads to the generation of a more accurate network for classification and a lower error rate. Second, the error rates are low for both the FNN and RNN in our detection system. For example, using a training set of 60 % with permissions-based detection, the error rate is 10 % using the FNN and

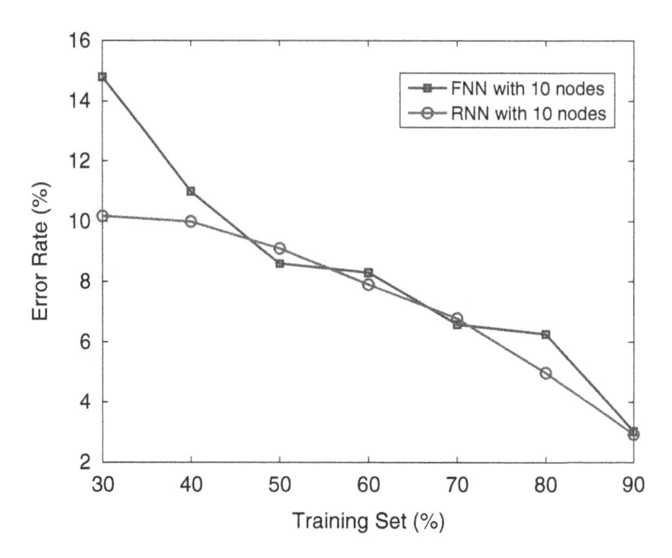

Fig. 7.14 Error rate for system call based detection vs. Training set ratio (1 g)

8 % using the RNN. Similar results have been obtained using system call based detection. Thus, we have confirmed that our detection system obtains high detection rates as well as low error rates, ensuring detection accuracy.

7.6 Discussions

In this section, we discuss some issues related to our malware detection system.

7.6.1 Overhead of Training Process

The major overhead of our ANN-based detection system comes from the training process. It is worth noting that the training process consists of procedures for collecting data sources, mapping data sources, and training the neural network. After the network is well trained, the online detection procedure can be fast. Overhead for the training process can be presented by $T = np(T_d + T_m) + T_l$, where n is the number of total applications, p is the training set ratio, and T_d, T_m, T_l are the average overhead for: dumping permissions and system calls from one application, mapping process and training the neural network, respectively.

As an example, consider training using 1-g. In our experiment, we implemented the permission-based detection and measured the execution time of each step. With $p = 90$ % and $n = 188$, the average time consists of 0.000343 s to dump permissions,

0.00012 s to map permissions, and 0.41 s to train neural networks. Hence, the total overhead of the training process is 0.613 s. Similarly, we investigated the overhead of system call-based detection. We note that in order to dump system calls associated with the execution of applications, we need to manually execute applications on real-world mobile devices and the execution times can be random, depending on the application. In our experiments, the overhead of mapping process is 0.00026 s and the total time for the training process is 0.194 s. It is worth noting that the computation overhead linearly increases with the number of applications. To make our system scale, one possible solution is to take advantage of powerful hardware for neuromorphic approaches to conduct threat analysis and detection.

7.6.2 Cloud Based Detection

We have developed an ANN-based malware detection system that detects unknown malware on mobile devices. Nonetheless, because a large number of mobile devices can be deployed in the system, those devices will generate big data associated with malware detection over time. Hence, mobile devices are characterized by limited storage capacity, constraint battery life time, and limited computational resources.

To address this issue, we shall investigate how to use the cloud computing infrastructure and algorithms to assist malware detection. Leveraging a cloud computing based service to store monitoring threat detection can expand resource and storage capacity and enhance the efficiency of threat analysis. By leveraging the cloud computing infrastructure, a monitoring agent can be deployed in the mobile device to collect permissions and system calls associated with applications then transmit such data sources to the remote cloud server. We can integrate our ANN-based detection and other detections schemes in the cloud server to help the human administer to defend against malware attacks.

7.7 Related Work

The detection of malware on a mobile platform can be categorized into static analysis, dynamic analysis, and permission analysis. These techniques have been investigated in the past by [12–16]. For example, Bose et al. [12] proposed a malware behavioral detection scheme on mobile handsets. Shamili et al. [13] presented a distributed Support Vector Machine (SVM) scheme to conduct malware detection, along with a statistical classification model. Deepak et al. [14] proposed a signature-based malware detection scheme. Schmidt et al. [15] conducted the static analysis of malware on the Android platform. To measure the effectiveness of different schemes on malware detection, Shabtai et al. [16] evaluated several classification and anomaly detection schemes and feature selection methods for mitigating malware on mobile devices.

Through permission analysis, malware detection can be conducted through the analysis of extracted security configurations and policy rules [10, 17–20]. For example, Aung et al. [18] developed a machine learning-based detection on the Android platform by monitoring permission related features and events. Huang et al. [20] conducted the permission-based detection for Android malware by using machine learning schemes such as AdaBoost, Naive Bayes, Decision Tree (C4.5), and Support Vector Machine. David et al. [19] presented a Self- Organizing Map (SOM) scheme to identify the permission-based security model using 1,100 android applications.

Neural networks can be used to learn and classify anomaly activities based on limited data sources [21]. There have been a number of research efforts on using neural networks to conduct threat detection [21–24]. For example, Mukkamala et al. [22] investigated schemes to conduct intrusion detection using neural networks and SVMs. Linda et al. [23] proposed a neural network-based approach to conduct intrusion detection for critical infrastructures. Golovko et al. [24] discussed the use of neural networks and artificial immune systems for carrying out malware and intrusion detection.

Different from existing research efforts, our detection system considers both permission and system calls as data sources. To learn the behavior of malware and benign applications, our system compares the performance of two classical neural networks: FNN and RNN. We have also shown that our implementation can detect unknown malware.

7.8 Conclusion

Malware attacks on smart mobile devices have been growing and posing security risks to mobile users. In this chapter, we developed an ANN-based malware detection system to automatically learn the behavior of applications and to detect unknown malware. In our developed system, we systematically compared the permission requests from application requests and system calls to capture the behavior of applications. Using real-world malware and benign applications, we conducted experiments on Android mobile devices. Our data shows the effectiveness of our developed detection system.

References

1. What is Android? http://android.pk/android.html.
2. Smartphones account for half of all mobile phones, dominate new phone purchases in the us. http://www.nielsen.com/us/en/newswire/2012/smartphones-account-for-half-of-all-mobile-phones-dominate-new-phone-purchases-inhtml.
3. A. Nere, A. Hashmi, M. Lipasti, and G. Tononi: Bridging the Semantic Gap: Emulating Biological Neuronal Behaviors with Simple Digital Neurons. In Proceedings of IEEE 19th International Symposium on High Perfor- mance Computer Architecture (HPCA), (2013).

4. D. J. Montana and L. Davis: Training Feedforward Neural Networks Using Ge- netic. In Proceedings of International Joint Conference on Artificial Intelligence Algorithms, (1989).
5. X. Yu, M. O. Efe, and O. Kaynak: A General Backpropagation Algorithm for Feedforward Neural Networks Learning. In IEEE Transactions on Neural Net- works, vol. 13, pp. 251-254 (2002).
6. G. Arulampalam and A. Bouzerdoum: A Generalized Feedforward Neural Network Architecture for Classification and Regression. In Journal of Neural Networks, vol. 16, pp. 561-568 (2003).
7. J. Y. F. Yam and T. W. S. Chow: A Weight Initialization Method for Improving Training Speed in Feedforward Neural Network. In Neurocomputing, vol. 30, pp. 219-232 (2000).
8. S. Kak: On Training Feedforward Neural Networks. In Pramana-Journal of Physics, vol. 40, pp. 35-42 (1993).
9. A. D. Schmidt, R. Bye, H. G. Schmidt, J. H. Clausen, O. Kiraz, K. Yuksel, S. A. Camtepe, and S. Albayrak: Static Analysis of Executables for Collaborative Malware Detection on Android. In Proceedings of the IEEE International Conference on Communications (ICC), (2009).
10. M. Grace, Y. Zhou, Z. Wang, and X. Jiang: Systematic Detection of Capability Leaks in Stock Android Smartphones. In Proceedings of the 19th Annual Symposium on Network and Distributed System Security (NDSS), (2012).
11. I. Burguera, U. Zurutuza, and S. Nadjm-Tehrani: Crowdroid: Behavior-based Mal- ware Detection System for Android. In Proceedings of the 1st ACM Workshop on Security and Privacy in Smartphones and Mobile Devices, (2011).
12. A. Bose, X. Hu, K. G. Shin, and T. Park: Behavioral Detection of Malware on Mobile Handsets. In Proceedings of the 6th ACM International Conference on Mobile Systems, Applications, and Services, (2008).
13. A. S. Shamili, C. Bauckhage, and T. Alpcan: Malware Detection on Mobile Devices using Distributed Machine Learning. In Proceedings of 20th IEEE International Conference on Pattern Recognition (ICPR), (2010).
14. D. Venugopal and G. Hu: Efficient Signature based Malware Detection on Mobile Devices. In Journal of Mobile Information Systems, vol. 4, no. 1, pp. 33- 49 (2008).
15. A. D. Schmidt, R. Bye, H. G. Schmidt, J. Clausen, O. Kiraz, K. A. Yuksel, S. A. Camtepe, and S. Albayrak: Static Analysis of Executables for Collaborative Malware Detection on Android. In Proceedings of IEEE International Conference on Communications (ICC), (2009).
16. A. Shabtai: Malware Detection on Mobile Devices. In Proceedings of the 11th IEEE International Conference on Mobile Data Management (MDM), pp. (2010).
17. A. Dinaburg, P. Royal, M. Sharif, and W. Lee: Ether: Malware Analysis via Hardware Virtualization Extensions. In Proceedings of the 15th ACM Conference on Computer and Communications Security (CCS), (2008).
18. Z. Aung and W. Zaw: Permission-Based Android Malware Detection. In International Journal of Scientific and Technology Research, vol. 2 (2013).
19. D. Barrera, H. G. Kayacik, P. C. van Oorschot, and A. Somayaji: A Methodology for Empirical Analysis of Permission-based Security Models and Its Application to Android. In Proceedings of the 17th ACM Conference on Computer and Communications Security (CCS), (2010).
20. C.-Y. Huang, Y.-T. Tsai, and C.-H. Hsu: Performance Evaluation on Permission-based Detection for Android Malware. In Springer Berlin Heidelberg, pp. 111-120 (2013).
21. J. Cannady: Artificial Neural Networks for Misuse Detection. In Proceedings of National Information Systems Security Conference, (1998).
22. S. Mukkamala, G. Janoski, and A. Sung: Intrusion Detection Using Neural Networks and Support Vector Machines. In Proceedings of IEEE International Joint Conference on Neural Networks, (2002).
23. O. Linda, T. Vollmer, and M. Manic: Neural Network based Intrusion Detection System for Critical Infrastructures. In Proceedings of IEEE International Joint Conference on Neural Networks, (2009).
24. V. Golovko, S. Bezobrazov, P. Kachurka, and L. Vaitsekhovich: Neural Network and Artificial Immune Systems for Malware and Network Intrusion Detection. In Advances in Machine Learning II. Springer, pp. 485-513 (2010).

Chapter 8
Sustainability Problems and a Novelty in the Concept of Energy

Simon Berkovich

8.1 Introduction

Energy is the major commodity for the industrial society. In general, energy is an ability to move matter, which in one way or another shows up universally. As physical processes unfold energy goes from one form to another, and the total sum of different energy values in commeasurable units remains constant (see [1]). This constitutes the law of conservation of energy—an important operational principle affirming that energy can neither come out of nowhere nor disappear into nothingness.

There are two ways for a resource to be supplied that can be specified as an "inflow" or as a "storage". For mechanical activities this relates to "kinetic" and "potential" energy. According to the worldview of the contemporary cosmology all the energy in the Universe had originated from the Big Bang being accumulated in the created matter in compliance with the famous formula $E=mc^2$. Thus, it is presumed that the primary energy of the Universe is a kind of a "potential" energy "stored" all over the matter.

To all appearance, energy contained in the matter is released as radiation from thermonuclear processes in the stars. One of these stars, the Sun, irradiates the Earth, and the "inflow" of the solar energy is the ultimate cause for all the processes in our world. For many millions of years the solar energy had been accumulated in various hydrocarbons serving as a chemical "storage" available for practical needs. Now, the reserves of hydrocarbons run down, and we face a serious crisis menacing our civilization. The situation is too well known to go into further detail.

S. Berkovich (✉)
Department of Computer Science, The George Washington University,
Washington, DC 20052, USA
e-mail: berkov@gwu.edu

R.E. Pino et al. (eds.), *Cybersecurity Systems for Human Cognition Augmentation*, Advances in Information Security 61,
DOI 10.1007/978-3-319-10374-7_8

There is a widespread opinion that mere innovations of the existing technologies, irrespective of the extent of their ingenuity, are not sufficient for an effective resolution of the energy crisis—"radical innovation, not incremental improvement is needed to make clean efficient energy" [2]. The pending calamities desperately call for a cardinal breakthrough in fundamental science. In this paper, we consider a suggestion for a new type of energy in association with the Holographic Universe [3]. Contrarily, to the belief in the primality of the "potential" energy $E = mc^2$ stored in the matter, the new surmised resource presents a "kinetic" energy—all-pervading continuous "inflow" of impetuses from a clocking mechanism, which is indispensable for the operability of the physical world. Impetuses from such a mechanism can be extracted with some process called "hot-clocking" [4] (Sect. 8.3).

The existence of a new type of energy could be suspected considering many observations where regular sources of energy are difficult to identify. In practical aspect, the most insistent situation presents the puzzling "excess heat" effect, which is attributed, although unconvincingly, to nuclear reactions. Lack of a solid scientific explanation is the major impediment on the way of reducing this promising effect to practice (Sects. 8.2 and 8.4).

Energy problems are of especial significance for the development of network systems. Having a dependable autonomous source of energy is decisive for functioning of vast distributed networks of sensors and actuators. Substantially, a dependable decentralized supply of energy would obviate physical vulnerability of the power grid.

8.2 Appearances of Unidentified Sources of Energy in Nature

There are many situations in Nature where the source of energy cannot be clearly determined. Attempts to interpret these situations using common knowledge are accompanied with a lot of doubts, or at least they do not go without hard questions. So, the possibility of involvement of a new type of energy cannot be excluded.

Typical physical effects with an unclear display of energy are delineated in Fig. 8.1. Obviously, the usual suspects for the appearing inconsistencies are nuclear reactions.

The most notorious example presents the "excess heat" effect attributed to "Cold Fusion" that grabbed public attention after a sensational introduction by Fleishman and Pons about 25 years ago. Information on this subject is copious and constantly growing (see, in particular, [5, 6]). In view of the purported practical implementations this effect deserves a special consideration as given in Sect. 8.4.

Sonoluminescence is the production of light from ultrasonic waves [7]. These waves cause intense oscillations and cavitations of bubbles in water producing extreme bursts of energy. The mechanism of this effect has not been explained despite significant experimental and theoretical studies, nuclear fusion may be suspected. Another strange outcome occurs in such a seemingly simple effect as

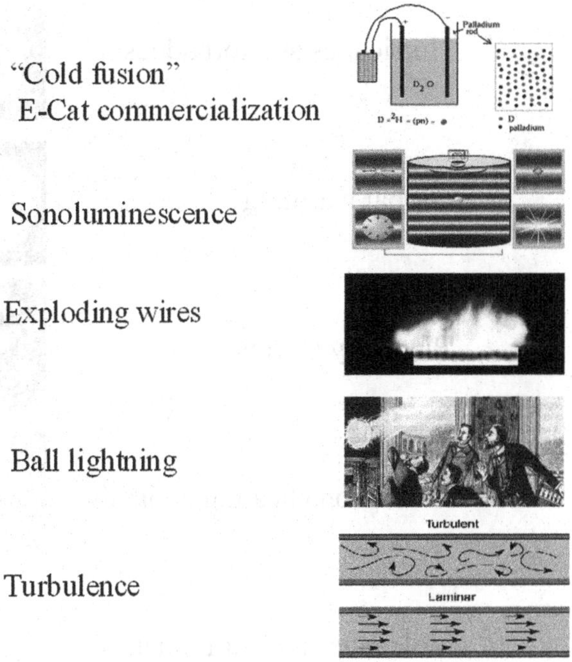

Fig. 8.1 Energy paradoxes in physical effects

"Cold fusion"
E-Cat commercialization

Sonoluminescence

Exploding wires

Ball lightning

Turbulence

blowing out a wire by an electric pulse like what happens in a fuse. This effect reveals an unexpected picture of demolition. Rather than being destroyed in a singular weak point the wire is fragmented in many distinct pieces. Beside explanations by traditional electrodynamics a connection of wire fragmentation with nuclear transmutations has been also suspected [8]. Here is appropriate to mention that electrical wires acquire energy not from the current inside but from the electromagnetic processes outside, so the scheme of getting "hot-clocking" energy in wire demolition by an electric pulse would be similar to the scheme of acquiring energy by parametric resonance in quantum entanglement as introduced in [3].

An unexplained atmospheric electrical phenomenon of ball lightning is usually associated with thunderstorms. This effect refers to rare erratic appearances of strange luminous objects of various sizes that after a while explode with a substantial vigor [9]. As accentuated by P.L. Kapitsa [10], the main problem is that the relatively long endurance of the ball lightning necessitates a continuous inflow of energy from outside, and he assumed that this energy should be supplied by microwave radiation. There are various other hypotheses, but at present time there is no acceptable explanation.

Turbulence, a chaotic flow regime, is an obscure phenomenon [11]. There is an apocryphal story about Werner Heisenberg: "When I meet God, I am going to ask him two questions: Why relativity? And why turbulence? I really believe he will have an answer for the first." According to Richard Feynman, turbulence is "the most important unsolved problem of classical physics." Energy processes in

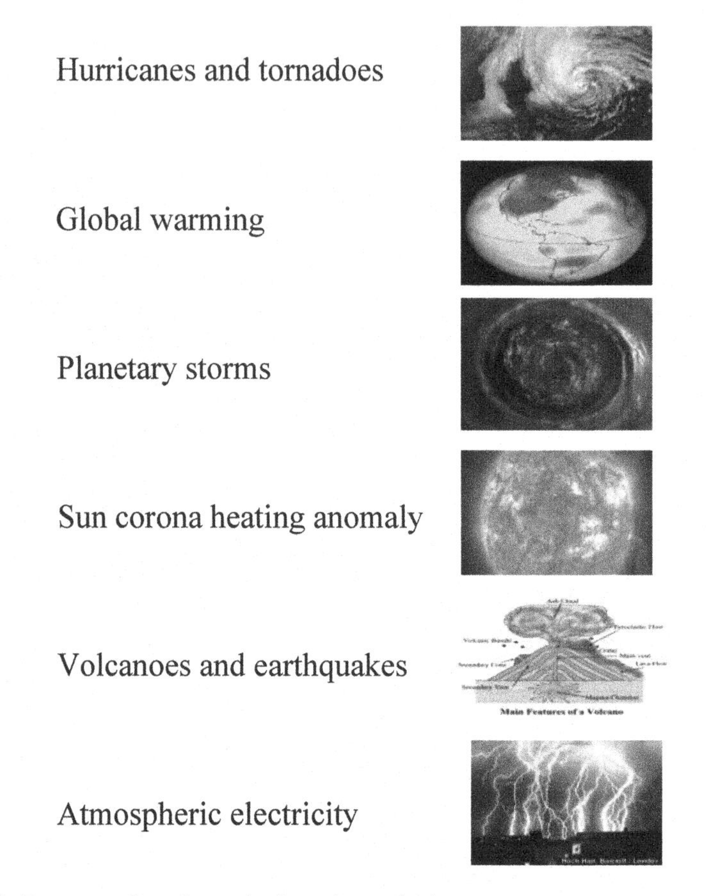

Fig. 8.2 Energy paradoxes in geophysics and astrophysics

turbulence are enigmatic. Momentously, these processes have great geophysical and astrophysical consequences.

Energy problems in relation to geophysics and astrophysics are put on view in Fig. 8.2.

The extreme weather events of hurricanes and tornadoes moving large masses of air require tremendous amounts of energy. Typical hurricane contains 10 billions kilowatt-hours (as a hydrogen bomb); certain tornadoes possess energy comparable to an atomic bomb. It is not exactly clear how such energy could be concentrated from solar radiation. It seems strange that well-determined mechanical motions, like vortices, could originate from a chaotic energy of heat. In our view, the surmised "hot-clocking" energy is what contributing to the heating and moving of atmospheric masses. These actions are enhanced in the air polluted with atmospheric aerosols. So, extreme weather events go simultaneously with warming up of the atmosphere. The possible supplement of the new energy can be estimated using ballpark figures from a recent EPRI report [12]. This report considers nanopowder for Ni-Pd alloys in the size of 5–10 nm that cause small amounts of "excess energy" (see Sect. 8.4), about 4 W per gram. A rough norm of air contamination is around 20 micrograms per cubic meter. So, the whole atmosphere could generate dozens of terawatts, a supplement comparable to the contribution from the Sun.

A remarkable situation presents atmospheric events at the planets of the solar system. Thus, tremendous storms have been observed at Saturn, which is much larger than Earthly hurricanes. They contain fastest winds—over 500 m per second. Saturn also has powerful lightning storms that are 10,000 times stronger than on Earth [13]. Considering energy required for these events one should take into account that the density of the solar radiation on Saturn is about 10 less than on Earth.

Not all is clear regarding solar energy. In June 2013, NASA launched a satellite to solve a longstanding solar mystery: why is the sun's upper atmosphere so much hotter than the sun's surface? [14]. The hottest temperatures of the sun about 15 millions degrees are supposed to be at its core, where fusion of hydrogen atoms occurs. The heat from the core reaches the surface, where the temperature is about 6,000°. Further on something bizarre happens: the temperature of the sun's corona rises to several million degrees. This does not make sense—temperatures usually drop as one moves away from a furnace.

"To explain this sudden jump in temperature, scientists have concluded that something in the region between the sun's surface and the corona, a thin area called the transitional region, is causing the corona to get superheated, but they are not yet sure what" [14]. A renowned Russian astrophysicist N.A. Kozyrev considered a curious possibility that star energy originates from "active flow" of time (see [15]). Our suggestion for energy generation by "hot-clocking" exposes a certain physical meaning of this idea.

High-energy events belong not only to the realm of hot stars. It turns out that such events are associated with thunderstorms in the Earth atmosphere including blasts of gamma rays, X-rays, beams of particles, even of antimatter. "The atmosphere is a stranger place than we ever imagined" [16]. Also, a number of high-energy uncertainties in relation to the internal structure of the Earth display its tectonic activities [17].

The issue of energy is a pivot for the sustenance of Life. Multifaceted questions of how energy impels living creatures do not have definite answers (Fig. 8.3). The number of muscles in the human body is about 650–850, and it is not clear how energy can be delivered with precision and timeliness. "The more data we collect regarding muscle the less we understand its functions" and "here we have approached a chasm going through the whole medicine and biology" [18].

The power density in conversion of chemical energy into mechanical work is about 200 W per m^2. Unrealistic situation—an animal producing one Horse Power (about ¾ Kw) must possess muscles with a surface of more than 3 m^2. So, muscles cannot get efficient energy flow through their surface from outside, the required flow of energy can arrive only through the volume from inside. A noteworthy observation had been reported in [19]: "mouth-rinsing and then spitting out a carbohydrate solution immediately improved performance at sprinting and cycling—even though it takes at least 10 min for carbohydrates to be digested and utilized by muscles".

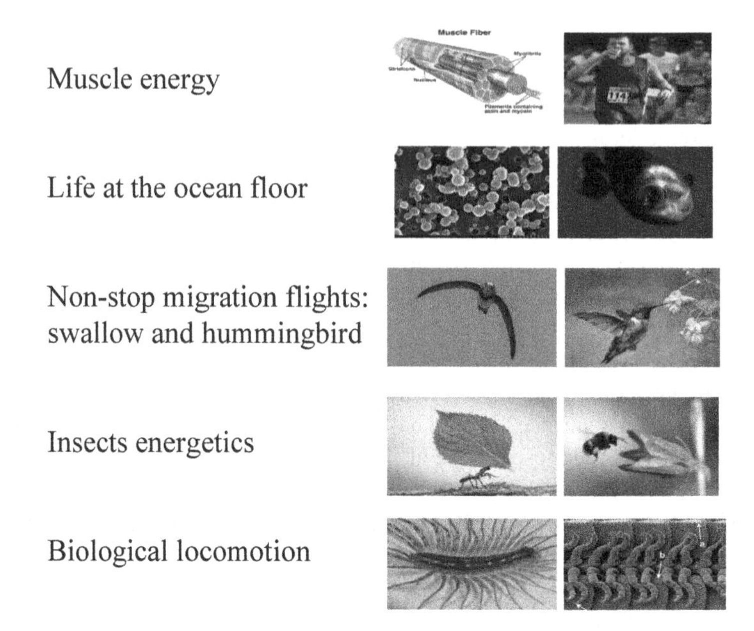

Fig. 8.3 Energy paradoxes in biological systems

Deep-sea organisms have no access to sunlight. So, they are believed to get their energy using as a food chemosynthetic bacteria developing from the hydrothermal vents [20]. Also, it is assumed that partially deep-sea organisms are dependent on a continuous shower of organic waist falling from the upper layers. Likewise, such a doubtful way of acquiring energy from food scattered in the surroundings is envisioned for "extraordinary behavior" of certain birds, like Alpine swifts that can fly thousands of miles without stopping. These "birds expend lots of energy during flight, but do not need to stop to eat because they feed midair on what is called aerial plankton—the atmospheric equivalent to marine plankton that can include an array of tiny bacteria, fungus, seeds, spores and small insects that get caught in air current" [21]. Hummingbirds also have another extremely energy consuming facility of hovering, as well as have many insects.

"In studies of insect energetics the completed budget rarely balances" [22]. Monarch butterflies are one of the few insects capable of making trans-Atlantic crossings. Some beetles would need daily intake of food twice their own mass. A bumblebee can travel 2,000 miles on the energy found in one teaspoon of nectar, but each flower supplies a portion only enough to keep the bee going for one minute. So, they must feed all day, stopping at over 100 flowers on each trip from the nest. Also, to examine the issue of energy balance in biological systems it could be instructive to compare the calorific values of massive trees with the amounts of solar energy pertained to their developments.

Another very unclear question is how energy could be continuously apportioned among the numerous parts of an organism, like in locomotion of 750-leg multipede [23].

8.3 The Origination of Energy from "Hot-Clocking"

The examples considered above show a number of vital phenomena where it is difficult to pinpoint the driving energy. Apparently, this suggests that Nature might contain a new yet not recognized source of energy. The introduced hypothesized energy has properties that are somehow different from those of ordinary sources.

Predominantly, it is believed that natural phenomena operate on latent facilities stored in the material formations as thermal, chemical, or nuclear energy. These facilities present what can be called "potential" energy. Other natural phenomena are propelled by fluxes of solar radiation, winds, torrents, and various kinds of streams. So, they employ what can be called "kinetic" energy. The new type of energy is of the latter, "kinetic", type. It comes from a continuous flow of certain omnipresent and seemingly inexhaustible impetuses, although they are imperceptible. Characteristically, as with all streams, at any given moment only a small portion of the incoming flow of energy results in effectual outcomes, the rest dissolves in unrelated disordered motions.

According to the existing worldview, the Big Bang at one single moment had provided the Universe with a definite amount of energy integrated in the created matter. So, it looks like Nature simply relies at one time initial imposition of a finite amount of energy without further replenishments. The given initial energy is not being destroyed; it just transforms from one type of accumulation to another. The quantity of this energy derived from matter is conserved, but its quality gradually degrades going from highly organized mechanical and electrical types of energy to less organized chemical and thermal types of energy. Eventually, the degradation of the quality of energy would lead to the paradox of "the heat death of the Universe".

Defiantly, in the surmised contraption the basic phenomena of Nature are continuously supplied with clean unlimited energy. The influx of this energy essentially supplements and revises the whirligig of energy transformations in the conventional picture of the Universe. A malevolent question immediately arises: what kind of a ubiquitous process can occur in Nature that misses the scrutiny of the researchers? In the framework of the Holographic Universe the surmised influx of energy is determined by the indispensable clock pulses in conjunction with the creation of periodic holography reference wave-trains. Clock pulses generation is a *conditio sine qua non* for any information-processing device. But asking a specialist what is the main part of a computer you hardly get the answer that it is the clock. Taken for granted, ubiquitous things escape people's attention.

Normally, a clock pulse generator serves as guide to carry out a given sequence of events, similarly to the role of an orchestra conductor. Yet besides merely issuing control signals a clock generator could provide also some tangible impetuses. Figuratively thinking, one can imagine that an orchestra conductor in addition to sending visual information messages can incite certain air movements to stimulate execution of the signals. There is a possibility to extract this way energy from clock pulses; it is called "hot-clocking" [4]. This idea is presented in Fig. 8.4. The scheme is called "hot-clocking" because "the voltage supply operates at the same time as

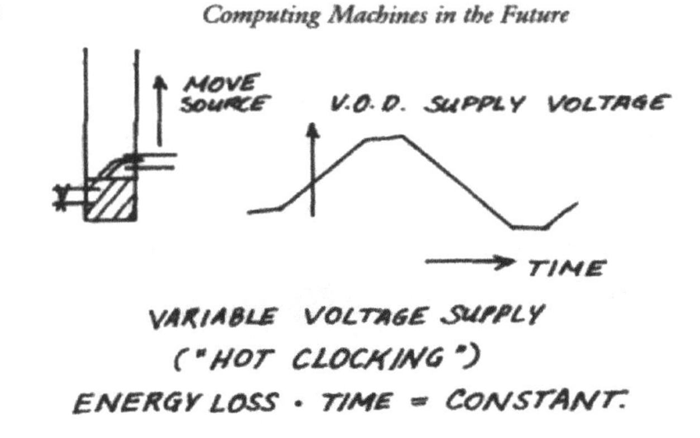

Fig. 8.4 Feynman's scheme of "hot-clocking"

the clock which times everything. In addition, we don't need an extra clock signal to time the circuits as we do in conventional designs" [4].

Combining in one channel information control and energy actuation is an effective common practice, like it is done in the USB ports of computers. The construction of the Holographic Universe combines quantum computing and quantum dynamics. A plain engineering illustration of this kind of a concept gives a device for harvesting energy from a beating heart, which collects enough energy to keep a pacemaker running [24].

According to Aristotle, the world is permeated with some global rhythm. The clocking of the Holographic Universe provides such a rhythm that can be responsible both for information control and energy actuation. This is the pivot point for the presentation of the physical world as an Internet of Things [25, 26].

8.4 Interpreting the "Excess Heat" Observations

Among the energy paradoxes in physics (Sect. 8.2) the situation with the "excess heat" appears most close to practical realization. The "excess heat" effect refers to observations of anomalous production of heat in certain hydrogen-metal systems. This effect creates a great controversy being firmly attributed to some nuclear processes dubbed "Cold Fusion" and later on Low Energy Nuclear Reactions (LENR) (see, e.g., [27]). Regretfully, long attempts to verify this effect have not resulted in an acceptable explanation. In recent years, the predominant attention has attracted the Energy-Catalyzer (E-Cat) device developed by Italian inventor Andrea Rossi (see, in particular, [28]). Our view of the problem is given in [29].

The E-cat device is rather simple. It contains, basically, a powder of nickel and gaseous hydrogen. Yet, it also utilizes some not specified additives: presumably, according to hearsay evidences and our theoretical speculations, those present a certain organic substance and microwave stimulation, as it interferes with "hot-clocking". Reportedly, E-Cat generates more energy than it consumes, by a factor of

6–30, and in a self-sustained mode even without any input at all. Latest experimental tests [28] have found energy densities in E-Cat "to be far above those of any known chemical source. Even by the most conservative assumptions as to the errors in the measurements, the result is still one order of magnitude greater than conventional energy sources." The tests were terminated by a deliberate shutdown, not by fuel exhaustion; next planned test experiments will last longer for further attempts to unveil the origin of the observed heat phenomenon.

It is commonly expected that new ideas be met with a resistance. Yet the "excess heat" effect, in some way, presents an extraordinary case. How could it happen that in our age of advanced technology an experimental verification of an apparently simple physical setup at the level of undergraduate physics—presence or absence of some extra heat—has not been reliably established despite the excruciating efforts about two decades? The question is not whether the effect does exist or does not. The question is why this problem of paramount importance cannot be clearly exposed.

Two exceptional circumstances are responsible for this confusing state of affairs. First, sticking to the paradigm of modern physics the appearance of substantial heat cannot be expected other than from a nuclear reaction. But the attempts to explain the production of the excessive heat in terms of "Cold Fusion" run into an irreconcilable conflict with the well-established body of physical knowledge. Nuclear fusion seems impossible at ordinary temperatures and pressures, primarily, because like-charged atomic nuclei do not have sufficient might to overcome the Coulomb barrier. Various modifications of this process suggest an approach that could make the low energy nuclear reactions (LENR) happen. Essentially, a hydrogen cation H^+—a proton—can capture an electron transforming into a neutron, then the uncharged neutron obviates Coulomb barrier and penetrates into a positively charged nucleus. The subsequent nuclear transmutation could release energy. However, it is generally agreed that different nuclear fusion adaptations do not explain the "excess heat" effect satisfactorily.

The "excess heat" observations do not reveal emission of particles that normally accompany nuclear reactions. Especially, gamma radiation could not pass unnoticed, as it would make a great harm to the experimenters' health. One can argue that emission of particles could be somehow shielded within the apparatus. But this is not the case for neutrinos that are not stopped by any obstruction. If no neutrinos were detected the "Cold Fusion" explanation of the "excess heat" should be completely ruled out. Presumably, these investigations are not considered worthwhile, and not only because they are costly and complicated. Uncomfortably, the "excess heat" experiments do not comply with the primary standard scientific requirement of reproducibility. So, the other circumstance that prevents enthusiastic acceptance of the "excess heat" effect is that it cannot be reliably reproduced. The effect is successfully detected only in approximately 30 % of experiments. This raises allegations that reports on this effect are results of professional incompetence and/or deliberate fraud (see, e.g., [30]).

However, the reasons for the irreproducibility of the considered phenomenon are much more profound. Here we encounter an absolutely new type of a physical phenomenon—not just in terms of its mechanism, but also in terms of its "statistical organization". Ordinarily, scientific method deals with values of measurements whose average yields the target parameter. Recently, it has been discovered that certain nanoscale phenomena with power-law distribution—like blinking light emitters—do not have an average value. An extended excerpt from the paper [31] describes the problem: "Imagine driving your car at night while its headlights display an annoying blinking behavior, switching on and off randomly. To add to the nuisance, the blinking has no definite time scale. In fact, although in most of your nightly journeys your headlights display quite rapid blinking, rendering at least some visibility, occasionally they remain off for almost the entire journey. Ridiculous and impractical as that behavior may seem, such is the situation commonly encountered by nanoscientists: A wide variety of natural and artificial nanoscopic light emitters, from fluorescent proteins to semiconductor nanostructures, display a blinking behavior like that described above. The emission (on) and no-emission (off) periods have a duration that varies from less than a millisecond to several minutes and more. The probability of occurrence of the on and off times is characterized by a power law, which is a typical sign of high complexity and is fundamentally different from what is expected from the quantum jump mechanism of fluorescence blinking predicted at the dawn of quantum mechanics."

An adequate explanation of the power-law blinking evades all the experimental and theoretical efforts. It is suspected that this kind of erratic behavior is universal for the mesoscopic world. In our view, the surmised new source of energy is determined by "hot-clocking" in the framework of the Holographic Universe that creates a sophisticated setup for the mesoscopic world. Thus, the "excess heat" effect could be also associated with a nanoscale mechanism yielding statistical distributions without an average value. In connection with the unsteadiness of the "excess heat" effect in the considered framework of the mesoscopic world it should be noted that this setup could cause a peculiar behavior of living systems at nanoscale level leading to the irreproducibility in bio-medical research, which is a worry of general concern [32].

In accordance with the developed concept the behavior of mesoscopic systems is sensitive to their positioning with respect to the holographic machinery of the Universe. Since these positionings change regularly with the rotation of the Earth around the Sun the physical and biological phenomena exhibit some annual periodicity called "calendar" effect [33]. So, the fluctuating outcomes of the "excess heat" could be affected by this calendar effect in conjunction with the absolute spatial orientation of the apparatus.

The ability to move matter by the "hot-clocking" effect can reveal itself as a manifestation of different types of energy: mechanical, thermal, electro-chemical, and, possibly, even nuclear. Depending on the organization of material configurations to which "hot-clocking" stimulation is applied it can produce ordered movements of mechanical energy or chaotic movements of thermal energy as presented in Fig. 8.5. Ordered impacts from "hot-clocking" bring about all biochemical activities. A mechanical engine can be built by simulating workings of a muscle [3].

The "hot-clocking" incitement of electronic structures could also result in producing electrical energy similarly to the process in photovoltaic cells. Thus, following "Cold Fusion" allusions, a device of powder diodes with a large surface

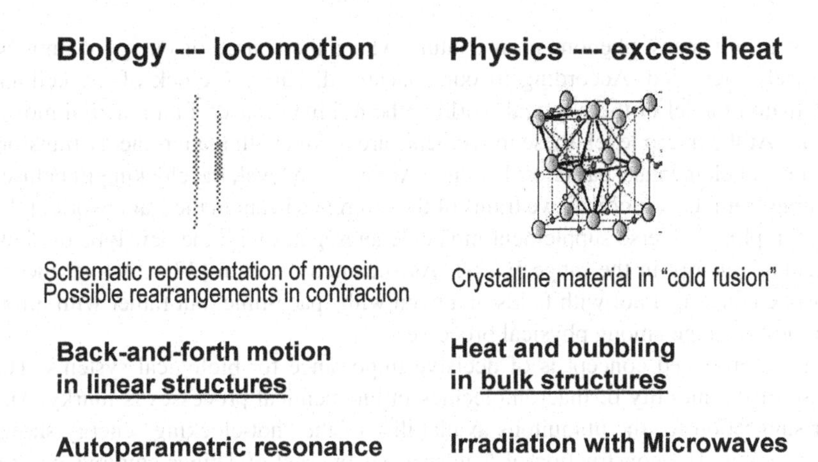

Fig. 8.5 Taking out "hot-clocking" stimuli in mechanical and thermal processes

area of semiconductor in contact with palladium and charged with deuterium was fabricated [34]. Reportedly, electrical energy comes out in a spontaneous potential difference of about 0.5 V.

8.5 Concluding Remarks

Behind any dynamic system there must be a supply of driving impulsions. For material systems of the physical world this implicates the concept of energy. The law of conservation of energy states that energy never appears out of nothing nor disappears without a trace: metaphorically speaking, it simply goes from one "formula" to another.

To utilize a sufficient amount of energy that is clean and easy available is crucial for the sustainability of human civilization. More than half a century aspirations for abundant energy from thermonuclear reactions are, mildly saying, not in the foreseeable future because of tough engineering limitations imposed by the severe operational requirements [35]. Purportedly, the anomalous "excess heat" effect claims the possibility of nuclear reactions in certain mild "cold fusion" conditions. Despite growing experimental support these claims do not raise enthusiasm in the scientific community essentially for two reasons: the interpretation of the "excess heat" effect contradicts the well established body of knowledge in nuclear physics, and this effect is not quite reproducible as the traditional scientific method mandates. In our view, what is considered "cold fusion" is in fact a "hot fusion" of matter, energy, and information—a new paradigm for fundamental physics leading to the most drastic reformation of the mesoscopic world. To begin with, we lay down a cellular automaton model of the physical world that generates the whole spectrum of elementary particles of matter; then, this model produces a holographic milieu framing the Universe as an Internet of Things [25, 26, 36].

There are many phenomena in Nature where the source of energy cannot be precisely identified. According to our concept, the driving clock of the cellular automaton model of the physical world is the ultimate cause of all material movements. At the micro level, these movements are linked with matter-energy transformations of elementary particles: $E=m\cdot c^2$. At the meso level, the clocking impetuses coming from the periodic wave trains of the reference beam in the framework of the Holographic Universe supplement molecule aggregates with the new type of "hot-clocking" energy in the range $E=k\cdot T$. As succinctly said in [37]: "A final theory must be concerned not with fields, not even with space-time, but rather with information exchange among physical processes".

The introduced concept is of decisive importance for biological systems. The cause of the motility of macromolecules in biochemical processes is murky. And our suggestion for the ubiquitous availability of the "hot-clocking" energy shows the way out. The construction of Nature as an Internet of Things employs control information from cloud computing and extracts actuation energy from "hot-clocking. Notably, the highly diversified supply of energy to neural componentry of the brain utilizes about 20 % of the energy delivered to the whole human body. According to the ideas of Erwin Schrödinger articulated in his famous book [38] living organisms acquire with food "negentropy" rather than energy. In other words, feeding of living organisms is "maintenance" not a "refueling". Crucially, the construction of the Universe as an Internet of Things with continuous inflow to biochemical processes of the control signals and the actuation energy is what allows living systems to withstand the otherwise inevitable thermodynamics degradation.

For the technology of large complex networks the suggested concept of energy generation can be important in two aspects. Above all, mimicking the organization of the biological systems would be a distinctive progressive application of the surmised energy source for the construction of wide-area networks of autonomous devices, like sensors, actuators, robots etc. Periodic replenishments of the energy stockpile of the elements of these networks are a hard problem. Regular methods of energy harvesting may not be sufficient for this purpose. Certain biological systems, like collections of bacteria and colonies of insects, present network formations that according to the "Internet of Things" paradigm utilize "cloud computing" information and "hot-clocking" energy. Artificial networks of autonomous devices spread over big geographic regions could be effectively developed in this way as well.

Further, the traditional thorny problem of how to protect the electrical power grid could be effectively resolved having a reliable independent source of energy as illustrated in Fig. 8.6. The decentralized supply of energy settles two equally difficult complications in relation to electrical grids—cyber security and physical vulnerability.

Current investigations of certain metal-hydrogen systems, like Rossi's E-Cat and some others [39], seem promising for generation of clean abundant energy. The main barrier for practical realization of these developments is a lack of consistent scientific understanding of the origin of the "excess heat" effect, whose explanations in terms of various hypothetical nuclear processes are not adequate. The possibility for extracting energy from the holographic mechanism of the physical world

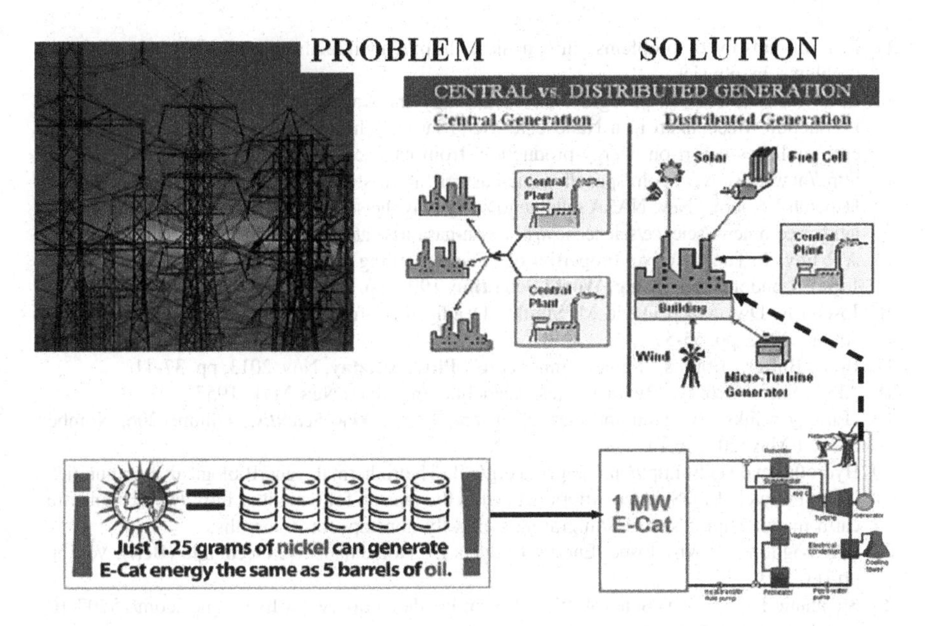

Fig. 8.6 Obviating problems of cybersecurity and vulnerability of the electrical grid with decentralized energy supply

by "hot-clicking" is in accord with the famous foreseeing of Nikola Tesla: "One day man will connect his apparatus to the very wheel work of the universe and the very forces that motivate the planets in their orbits and cause them to rotate will rotate his own machinery".

References

1. Richard Feynman, "Six Easy Pieces", Addison-Wesley Publishing Company, 1996
2. Mark Fischetti, "In Search of the Radical Solution", ***Scientific American***, Vol. 304, No 1, Jan. 2011, pp.52-55
3. Simon Berkovich, "Generation of clean energy by applying parametric resonance to quantum nonlocality clocking", **Nanotech** 2011, Vol. 1, pp.771-774
4. Richard Feynman, "The Pleasure of Finding Things Out", Perseus Publishing, Cambridge, Massachusetts, 2000
5. Cold Fusion Times, http://www.std.com/~mica/cft.html
6. E-Cat World, Following the Low Energy Nuclear Reaction Revolution, http://www.e-catworld.com/
7. Sonoluminescence, http://en.wikipedia.org/wiki/Sonoluminescence
8. A. Widom, Y.N. Srivastava, and L. Larsen, "Energetic Electrons and Nuclear Transmutations in Exploding Wires", January 01, 2007 http://iris.lib.neu.edu/cgi/viewcontent.cgi?article=1173&context=physics_fac_pubs
9. Ball lightning, http://en.wikipedia.org/wiki/Ball_lightning
10. P. L. Kapitsa, "On the Nature of Ball Lightning", In: "Experiment, Theory, Practice", Publishing House "Science", Moscow, 1987, pp. 55-61 (in Russian) Also, in Donald J. Ritchie. *Ball Lightning: A Collection of Soviet Research in English Translation* (1961 ed.)

11. List of unsolved problems in physics, http://en.wikipedia.org/wiki/List_of_unsolved_problems_in_physics
12. EPRI Institute Report, "Program on Technology Innovation: Assessment of Novel Energy Production Mechanism in a Nanoscale Metal lattice", http://www.e-catworld.com/2012/09/epri-publishes-report-on-energy-production- from-nanoscale-metal-lattice/
13. http://news.discovery.com/space/big-pic-cassini-saturn-storm-121128.html
14. Deborah Netburn, "New NASA satellite to study why the sun is so, you know, hot", http://www.latimes.com/news/science/sciencenow/la-sci-sn-nasa-iris-satellite- sun-20130626,0,6663891.story
15. A.P. Levich, The "'Active' Properties of Time According to N. A. Kozyrev", Singapore, New Jersey, London, Hong Kong, **World Scientific**, 1996, pp. 1- 42.
16. Joseph R. Dwyer and David M. Smith, "Deadly Rays from Clouds", **Scientific American**, August 2012, pp. 55-59
17. Bruce Buffet, "Erath's enigmatic inner core", **Physics today**, Nov. 2013, pp. 37-41
18. Albert Szent-Györgyi, Bioenergetics, Academic Press Inc., New York, 1957
19. "Energy drinks give your muscles an instant boost", ***New Scientist***, Volume 206, Number 2758, 1 May, 2010, p.14
20. Hydrothermal vents, http://en.wikipedia.org/wiki/Hydrothermal_vent#Biological_communities
21. Laura Poppick, LiveScience, Migrating swifts fly nonstop for 6 months, http://www.mnn.com/earth-matters/animals/stories/migrating-swifts-fly-nonstop- for- 6-months
22. J.A. Wightman, "Why Insect Energy Budgets Do Not Balance?", **Oecologia** (Berl), Vol. 50, pp.166-169, 1981
23. Stephanie Pappas, LiveScience, "750-leg millipede", http://www.livescience.com/25707-10-weirdest-animal-discoveries.html
24. "Heartbeat 'could power pacemaker'", http://www.bbc.co.uk/news/health-20182529
25. Simom Berkovich, "Physical world as an Internet of Things", COM.Geo'11, Proceedings of the 2nd International Conference on Computing for Geospatial Research and Application, Article No. 66, **ACM**, NewYork, NY, 2011
26. Nima Bari, Ganapathy Mani, and Simon Berkovich, "Internet of Things as a Methodological Concept", Computing for Geospatial Research and Applications (COM.Geo), 2013 Fourth International Conference,San Jose, CA, USA, July 22-24, 2013, pp. 48-55
27. Jed Rothwell, "Cold Fusion and the Future" http://lenr-canr.org/acrobat/RothwellJcoldfusiona.pdf
28. Giuseppe Levi, Evelyn Foschi, Torbjörn Hartman, Bo Höistad, Roland Pettersson, Lars Tegnér, and Hanno Essén, "Indication of anomalous heat energy production in a reactor device containing hydrogen loaded nickel powder", http://arxiv.org/ftp/arxiv/papers/1305/1305.3913.pdf
29. Simon Berkovich, New Physics of "Hot-Clocking Energy" for the "Excess Heat" Attributed to "Cold Fusion", http://www.bestthinking.com/articles/energy/new-physics-of-hot-clocking-energy-for-the-excess-heat-attributed-to-cold-fusion
30. Forbes' Gibbs on E-Cat Fraud Claims, http://www.e-catworld.com/2013/05/forbes-gibbs-on-e-cat-fraud-claims
31. Fernando D. Stefani, Jacob P. Hoogenboom, and Eli Barkai, "Beyond quantum jumps: Blinking nanoscale light emitters", ***Physics Today***, Feb. 2009, pp. 34-39
32. Elizabeth Iorns, "Is medical science built on shaky foundations?" **New Scientist**, 17 September 2012, issue 2882, pp.24-25
33. Simon Berkovich, "Calendar variations in the phenomena of Nature and the apparition of two Higgs bosons", http://www.seas.gwu.edu/~berkov/Berkovich_Calendar_Effect_modified.pdf http://www.bestthinking.com/articles/science/physics/quantum_physics/calendar-variations-in-the-phenomena-of-nature-and-the-apparition-of-two-higgs-bosons
34. Fabrice David and John Giles, "Self-Polarisation of Fusion Diodes: From Excess Energy to Energy", http://coldfusionnow.org/wp-content/uploads/2013/07/DavidFselfpolari.pdf
35. Michael Moyer, "Fusion's False Dawn", ***Scientific American,*** March 2010, pp.50-57

36. Simon Berkovich and Hanan Al Shargi, "Constructive Approach to Fundamental Science", **University Publishers,** San Diego, CA, 2010
37. Lee Smolin, "The Trouble with Physics", Houghton Mifflin Company, New York, NY, 2006
38. Erwin Schrödinger, "What is Life? The Physical Aspect of the Living Cell", Cambridge University Press, NY, 1992
39. Brillouin Energy Corporation, http://brillouinenergy.com/

Chapter 9
Memristors as Synapses in Artificial Neural Networks: Biomimicry Beyond Weight Change

Andrew J. Lohn, Patrick R. Mickel, James B. Aimone, Erik P. Debenedictis, and Matthew J. Marinella

9.1 Introduction

Cyberthreat security is a rapidly evolving landscape, where the diversity and number of attacks is constantly changing, requiring new approaches to defense. In the past, it was sufficient to predict likely attack methods and to monitor potential vulnerabilities, however, today the attacks are too varied and change too quickly for the traditional defenses to be effective. We need structures that are capable of identifying probable attacks and responding without human intervention. Due to the increasing rate of new attack methodologies, these structures need to be able to identify and respond to attacks that have never been seen before. That is, we need structures which are capable of learning.

Neural network implementations of computational machines were among the first to be studied [1] but they lost favor due to the simplicity and rapid progress in digital computers. As the computational power of these digital systems has grown it has become possible to implement software-based artificial neural networks using standard digital computation. However, when compared to the capabilities of the brain, it is clear that these approaches are far from reaching the full potential of neural networks or biologically inspired computation. Digital computers are capable of trillions of floating point operations per second (FLOPs) and storing immense information densities. They are capable of outperforming the human mind in a number of repetitive, deterministic, and well-specified problems, but are vastly inferior in aspects such as learning, adaptation and power consumption. In more complex tasks, the brain is able to outperform digital computation despite much lower speed because of the specialization of the neurobiological circuitry. We are starting to

A.J. Lohn (✉) • P.R. Mickel • J.B. Aimone • E.P. Debenedictis • M.J. Marinella
Cybersecurity Systems for Human Cognition Augmentation,
Sandia National Laboratories, Albuquerque, New Mexico 87185, USA
e-mail: drewlohn@gmail.com; prmicke@sandia.gov

R.E. Pino et al. (eds.), *Cybersecurity Systems for Human Cognition Augmentation*, Advances in Information Security 61,
DOI 10.1007/978-3-319-10374-7_9

understand how manufacturable electronic components can replicate some of these complex neurobiological phenomena and potentially use them to create efficient and powerful artificial neural networks that can adapt in real-time to a rapidly changing array of attack methodologies.

9.2 Memristors

Memristors are passive circuit elements that have recently revived substantial interest in artificial neural network approaches [2] to computing and there is increasing confidence that hardware-based neural networks can provide superior performance for a range of problems as compared to traditional digital software architectures. Theorized in the 1970s by Leon Chua [3, 4], memristors are resistors that are able to change their electrical resistance over time in response to stimulus in the form of voltage or current. When they were proposed by Chua, no physical embodiment of such a device was known but, in retrospect, their behavior had been observed in many systems for centuries [5] and had not been recognized as memristors. It was not until 2008 when a group at HP Labs led by R. Stanley Williams characterized the resistive switching in titanium oxide thin films as memristive switching [6] that their existence became widely known and studied. Since then, memristor research has grown rapidly and a vast range of new materials and devices have been identified as memristors, with many utilizing oxides [7–9], nitrides [10, 11], and chalcogenides [12–14] as the switching materials. There has been accelerated growth since Chua generalized his memristor formalism to show that any device with a pinched current voltage loop (see Fig. 9.1) can be characterized as a memristor [15]. Memristor development has also benefited from the technological relevance of

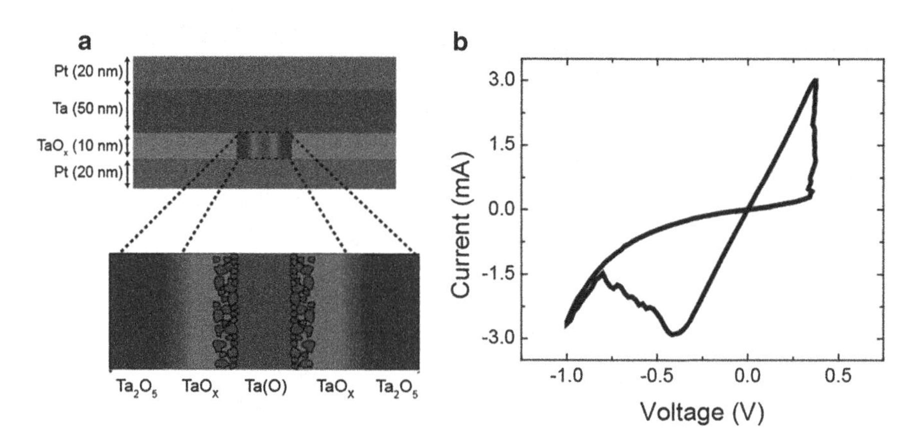

Fig. 9.1 A typical memristor architecture is shown where an insulating layer (TaO_x, here) is sandwiched between an inert (Pt) and reactive (Ta) electrode. Upon operation of the device a small conducting filament is formed, which we magnify here for clarity. (**b**) A typical current-voltage hysteresis loop is shown, where the device is reversibly switched between an Ohmic ON state and a non-linear insulating OFF state

resistive random access memory (ReRAM) [16, 17] (a form of memristor) which is one of the most viable candidates to replace FLASH memory as the dominant non-volatile memory device [18]. This has provided additional economic incentives as well as an extensive semiconductor research infrastructure for memristor technology. This is particularly true for a few specific materials including tantalum oxide (TaO_x) [19–21] and hafnium oxide (HfO_x) [22, 23] based devices where the oxide is sandwiched between a reactive and an inert electrode.

9.3 Memristors as Biological Synapses

The case for using memristors as artificial synapses for hardware-based neural networks has been relatively straightforward. Memristors change their electrical resistance state as a result of previous voltage differences across the device, which is directly analogous to synaptic weights that evolve between connected neurons. This synaptic weight modulation is at the heart of Hebbian learning, which is perhaps the simplest and most fundamental description of learning in neurobiological systems. Donald Hebb postulated in 1949 [24] when repeated activity occurs between cells, the efficiency of activity between those cells is increased. The resistance decrease of a memristor is therefore an obvious analog for Hebbian learning in artificial neural circuits.

This Hebbian approach to learning provides a justification for long term potentiation (LTP) observed in biological neural networks, where the synaptic weight is increased in a permanent or semi-permanent manner in response to excitation. Importantly, synapses are also known to exhibit long term depression (LTD) where the synaptic weight is decreased in response to excitation. Often, learning involves controlled increases and decreases of synaptic weights across a neural circuit, thus it is convenient that memristors can naturally exhibit both LTP and LTD-like behavior. By controlling the polarity of the applied voltage difference across the device the resistance state may be selectively increased (LTD) or decreased (LTP), with long time-scale retention. This simple view of memristors has made a very compelling case for their adoption as synapses in hardware-based neural circuitry and they have already proven successful in several scenarios [2, 25, 26], however these capabilities and successes still fall short of replicating even simple neurobiological systems. Creating a hardware-based system with more brain-like functionality will require an improved understanding of how memristors can be more accurately leveraged to emulate neurobiological systems.

9.3.1 Spiking Networks

One of the important differences between the approach to learning typically used in memristor-based networks discussed above and the approach to learning typically associated with neurobiological models is the existence of spiking. Most memristor networks implement classic artificial neural networks (ANNs) with perceptron-like neurons. Such perceptron networks are not typically dynamic in the way that neurobiological systems are; rather, the outputs of each individual node is updated each time step by a transformation of their inputs through a typically simple transfer function (i.e., a step function, sigmoid, or threshold-linear filter). The most typical approach for learning in these ANNs is simple backpropagation, where a set of inputs is provided and the generated output is compared to a known correct output. Then the weight of each synapse is adjusted in proportion to the difference between the known and generated outputs. It is worth noting that there are numerous other approaches to ANN training with different strengths and weaknesses, some of which are "supervised" like backpropagation (i.e., training guided by desired outputs), and some present a more Hebbian-like unsupervised learning (i.e., training guided by correlations of inputs alone). While it is widely recognized that the brain's approach is more complicated, it is worth noting that the major difference lies in the dynamic nature of neurons' behavior, in particular, their spiking function in which pulses of electrical activity propagate from one neuron to another.

Signals travel along nerve cells in spikes or pulses that are generated in response to a stimulus surpassing a threshold value. Early models to describe this behavior include integrate-and-fire [27] and the Hodgkin–Huxley model [28]. The integrate-and-fire model sums the inputs to the neuron and releases a burst at the output if the sum of the inputs is greater than a threshold. It does not output a signal if the sum of inputs is less than that threshold. The Hodgkin–Huxley model is slightly more complicated in that it attempts to model specific biological components with electrical circuitry. The membrane potential is calculated from parallel current paths corresponding to the sodium channels, the potassium channels, a membrane capacitance, and a current leakage path. This equivalent circuit model is well known to generate spikes that propagate through the neuron. A biologically-inspired artificial neuron should display similar behavior if it hopes to achieve brain-like functionality and memristors have been able to reproduce similar behavior.

In an analogous system, using memristors and capacitors to model ion channels that are energized with voltage sources, it was possible to create spiking behavior with features very similar to those observed in neurological cells [29]. This "neuristor," cell was able to demonstrate all-or-nothing spiking, several different biologically relevant spiking modes and spike trains, as well as a refractory period during which the neuristor (similarly to a neuron) must pause and recover prior to being able to spike again.

9.3.2 Spike Timing Dependent Plasticity

One of the ways in which spiking is particularly relevant is in the way that it influences LTP or LTD. A large body of experimental evidence [30–32] shows that often the time between spikes, as opposed to the spike magnitude, is the factor which determines the weight change of a synapse. In a simple illustration, when a spike arrives at the pre-synaptic neuron prior to the post-synaptic neuron, the synapse

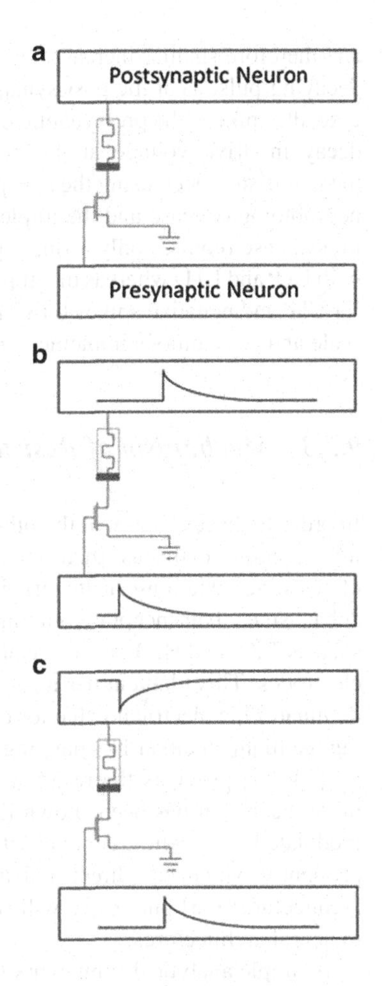

Fig. 9.2 A presynaptic neuron can be connected to a postsynaptic neuron by a memristor and a transistor as shown in (**a**) to mimic spike timing dependent plasticity. A decaying spike at the presynaptic neuron followed by a spike at the postsynaptic neuron (**b**) or postsynaptic spikes followed by presynaptic spikes (**c**) will produce LTP or LTD respectively

experiences LTP and when the spikes arrive in reverse order the synapse experiences LTD. The magnitude of the change in efficacy decreases as the time between the spikes increases. This effect is known as spike timing dependent plasticity (STDP) and efforts are being made toward reproducing it using memristors.

Using a simple circuit design that implements a silicon MOSFET with a HfO_x based memristor as the synapse it is possible to demonstrate STDP in memristor-based artificial synapses [33]. The implementation is structured with the pre-synaptic neuron wired to the gate of the transistor which has its source and drain wired in series with the memristor to the post-synaptic neuron as shown in Fig. 9.2.

When the pre-synaptic neuron is activated by a decaying voltage pulse, the MOS transistor is activated which allows a signal at the post-synaptic neuron to travel through the memristor and change its state (Fig. 9.2b). As the pulse at the pre-synaptic neuron decreases in magnitude, the degree to which the MOS transistor allows current to flow is reduced. In that way, increases in the timing delay between the pre and post-synaptic spikes lead to decreasing currents through the memristor and therefore smaller increases in synaptic weight. In the LTD case (Fig. 9.2c), the decaying pulse is at the post-synaptic neuron and has the opposite polarity. In this case, the spike at the pre-synaptic neuron acts to activate the MOS transistor and the decay in drive voltage at the post-synaptic neuron limits the current through the memristor, decreasing the synaptic weight. An important difference between the neurobiological case and the implementation described above is that the neurobiological case requires only a single polarity at the pre and post-synaptic neurons for both LTP and LTD whereas the implementation described above uses positive spikes for LTP and negative spikes for LTD. This difference is substantial and makes large scale and autonomous implementation as in the neuro-biological system difficult.

9.3.3 Mechanism of Resistance Change

In order to discuss many of the other biologically-inspired behaviors of memristors it is necessary to discuss their operation mechanism in more detail. The mechanism of resistive switching in memristive devices has been an active and unresolved debate since their inception but much has become clear. In the oxide-based devices such as TaO_x and HfO_x, a nanoscale filament is formed [34, 35] that bridges the two electrodes. This filament has a lower resistance than the surrounding material so it dominates the electrical behavior of the device. It has been proposed that modulating the filament either by changing its radius [19, 36] or by changing its conductivity [34, 37] provides the resistive switching effect that leads to memristance but more recently it has been shown [38] that in fact *both* radius and conductivity are modulated. The radius, conductivity, and filament shape can all be controlled independently which has direct and important implications in digital computational architectures [38] but, as we will see, also has a number of important implications in neural architectures.

A simple analytical framework [38] has been provided based on heat generation and dissipation in the filament. Their solution to the heat equation can be written so as to allow either the radius to change or the conductivity (i.e. the concentration of oxygen vacancies) to change. Provided that sufficient power is supplied to cause resistance change, the power of the electrical signal is related to the new resistance in the two cases (radius change or conductivity change) below:

$$IV_r = A_r \frac{T_{crit} - T_{RT}}{R - R_{min}} \tag{9.1a}$$

$$IV_\sigma = A_\sigma \frac{T_{crit} - T_{RT}}{R_{max} - R} \tag{9.1b}$$

where I is the electrical current, V is the voltage and the subscripts indicate whether radius (r) or conductivity (σ) are free to change. The resistance is R, the critical temperature for activating ion motion is T_{crit} and the ambient temperature is T_{RT}.

The remaining parameters: A_r, A_σ, R_{min}, and R_{max} are constants depending on materials and device design as described below:

$$A_r = \frac{2k_E d_O}{\sigma_{\max} d_E} \tag{9.1c}$$

$$A_\sigma = \frac{8d_O^2 L_{WF} T_{crit}}{r_{\max}^2} \tag{9.1d}$$

$$R_{\min} = \frac{k_E}{4\pi\sigma_{\max}^2 L_{WF} T_{crit} d_E} \tag{9.1e}$$

$$R_{\max} = \frac{4d_O^2 L_{WF} T_{crit} d_E}{\pi r_{\max}^4 k_E} \tag{9.1f}$$

where k_E is the thermal conductivity of the electrode, d_O and d_E are the thicknesses of the oxide and electrode respectively, L_{WF} is the Wiedemann–Franz constant, σ_{max} is the saturation conductivity that cannot be surpassed, and r_{max} is the maximum radius that the device has reached in its history.

Treating Eq. 9.1c–f as design factors that do not change during operation of the device, we can limit our discussion to Eqs. 9.1a and 9.1b which are really quite simple. They state that the electrical power is inversely proportional to the resistance state that is achieved, scaled by some factor and shifted by some amount. The radius change equation (Eq. 9.1a) is valid for decreasing resistance once oxygen vacancy saturation is achieved (i.e. positive voltage on the reactive electrode). The conductivity change equation (Eq. 9.1b) is valid for increases in resistance and for decreasing resistance until the point where oxygen vacancies saturate (i.e. negative voltage on the reactive electrode or positive voltage until saturation). By using the two different polarities for applied power it is possible to separately control the radius and conductivity of the filament.

Figure 9.3 describes the step-by-step progression throughout a switching cycle starting from a very high resistance state. To decrease resistance, one applies positive current to the reactive electrode (inert electrode grounded) and, once sufficient power is delivered, the conductivity starts to increase within a localized region to form a filament as shown in Fig. 9.3a. The conductivity continues to increase until it reaches saturation. Once saturated, and with continued positive current, the radius of the filament increases as shown in Fig. 9.3d. Now, starting from a saturated filament of a given radius, negative voltage can be applied to the reactive electrode (inert electrode grounded) to increase the resistance. This process occurs by reducing the conductivity of the filament and can be thought to proceed quasi-uniformly throughout the filament as shown in Fig. 9.3g.

The resistance decrease traces out a path to the left of the power-resistance curve (Fig. 9.3b,e) and the resistance increase traces out a path to the right (Fig. 9.3h). Since the resistance decrease due to radius increase leads to increases in power

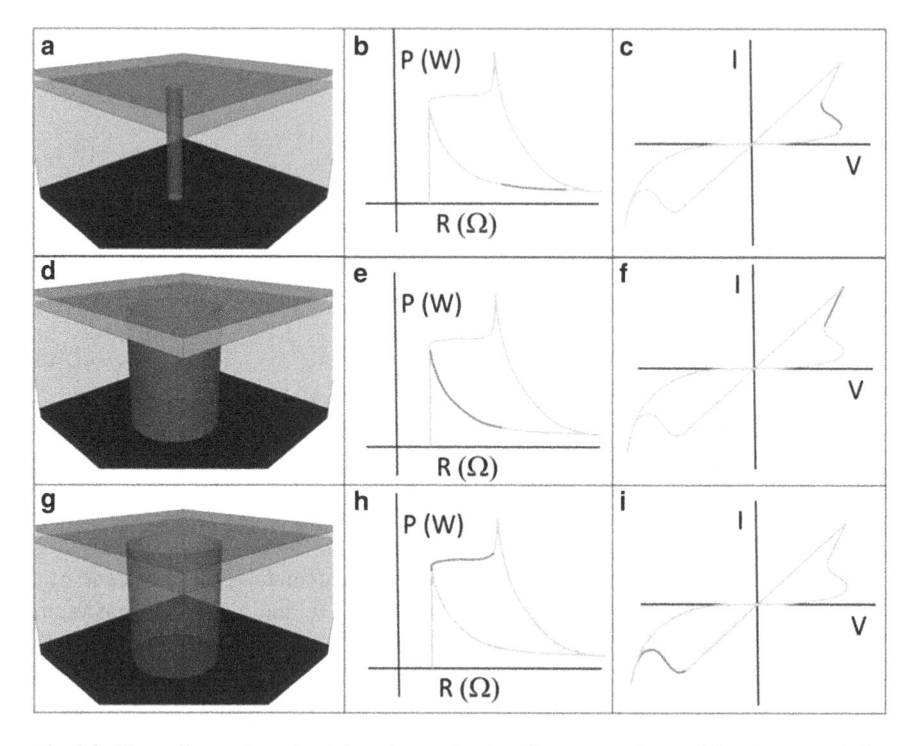

Fig. 9.3 The radius and conductivity of a conducting filament can be modulated separately by electrical means. Starting from a resistive state, positive polarity power (inert electrode grounded) creates a new filament (**a–c**). Once vacancy concentration reaches saturation the radius of the filament increases (**d–f**) and by changing the polarity of the power the conductivity can be decreased (**g–i**)

required for state change, a range of power resistance combinations exist. Furthermore, since the resistance increase due to conductivity decrease progresses toward the right in the power-resistance curve, starting from the last point in the previous radius change curve, the state of the device can be placed anywhere within the bounds indicated on the power-resistance plot [39]. Repeating the cycle, starting from a partially depleted filament, the positive current will create a new filament within the partially depleted one that has its own radius and conductivity which can be modulated as well [40].

9.3.4 Morphological Change

In discussing the Hebbian learning models previously, the effect of the rate of LTP or LTD was omitted. Those rates determine the learning rate and also the forgetting rate, both of which are critically important in a neural system. However, in attempting to attain both of these computational features, one is confronted with the stability-plasticity dilemma. It is desirable for a system to learn new things rapidly, which involves both LTP and LTD, but it is also desirable to keep an effective solution unchanged. In terms of the rates of LTP and LTD, the two are at odds. Rapid learning requires a high degree of plasticity and stability requires low plasticity.

One way by which the neurobiological system may address this problem is by adjusting the morphology of the synapses [41, 42]. Morphological changes in biological synapses [43, 44] are a complex process which is an ongoing topic of active research. However, in this biomimetic context a simplified model will be sufficient. In the synapse there are a number of receptors which open ion channels (leading to current flow into or out of the synapse) when triggered by a neurotransmitter. Broadly speaking, a higher number of these receptors at the synapse contributes to a higher synaptic weight and their density can be modulated to change the signaling efficiency. These receptors are usually introduced in a transient way which provides an increase in efficiency that degrades as the receptors are lost over short periods of time. These changes can be made more permanent through further signaling involving gene expression. The gene transcription and protein synthesis that results in increased stability is often associated with an increase in the area of the synapse and so morphological increases in size are often linked with increased stability [45]. This is one of the ways that biological synapses are able to overcome the stability-plasticity dilemma to maintain the capacity for both rapid learning and stable expertise.

Considering the resistive switching mechanisms discussed in Sect. 9.3.3, the same synaptic development can be manifested in filamentary memristive synapses. The electrical resistance (inverse of synaptic weight) of the memristive filament can be modulated by changing the density of current-carrying sites (see conductivity modulation, Eq. 9.1b) or by changing the area of the filament (see radius modulation, Eq. 9.1a). Just as in the biological case, increasing the area of the filament acts to stabilize the state of the synapse against small signals. Physically, this results from the fact that resistive switching is thermally activated, and the increased filament-electrode contact area allows heat to escape quickly, requiring additional Joule heating to reach the critical thermal activation temperature to initiate resistive switching.

This effect can be used to address the stability-plasticity dilemma at the synapse level in resistive-switching based synapses. Figure 9.4 shows a power-resistance plot for a filament in two different configurations corresponding to the same synaptic weight (i.e. resistance). In the first configuration, the filament is set with a small radius and large conductivity, while in the second configuration the filament is set to a large radius and small conductivity. As described above, because of the modified heat flow between the configurations, the second configuration requires substantially more applied power to thermally activate resistive switching. As shown in Fig. 9.4, a signal with at least 1 mW of peak power would be needed to modify configuration 1, whereas a signal with at least 1.75 mW of peak power would be needed to influence configuration 2.

In a biological synapse, transient changes in efficacy are made more permanent in response to a chemical signal indicating a positive outcome. An analogous signal

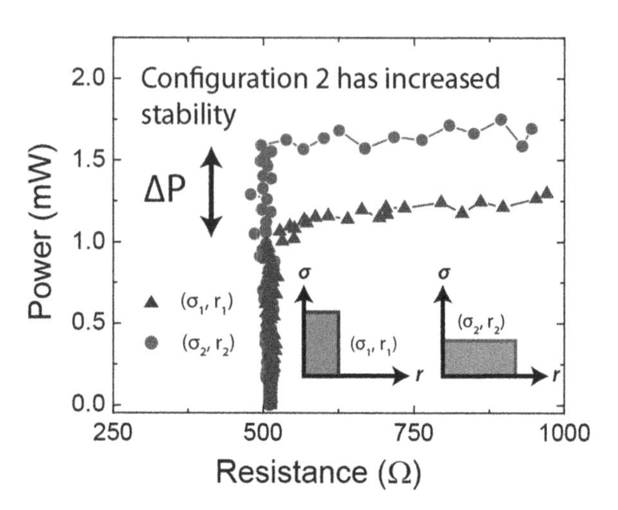

Fig. 9.4 Resistive switching power thresholds are shown for two distinct filament configurations: small-radius/large-conductivity and large-radius/small-conductivity. As seen, the large-radius/small-conductivity filament configuration requires as much as 75 % more applied power to modify the synaptic weight, and is substantially less plastic

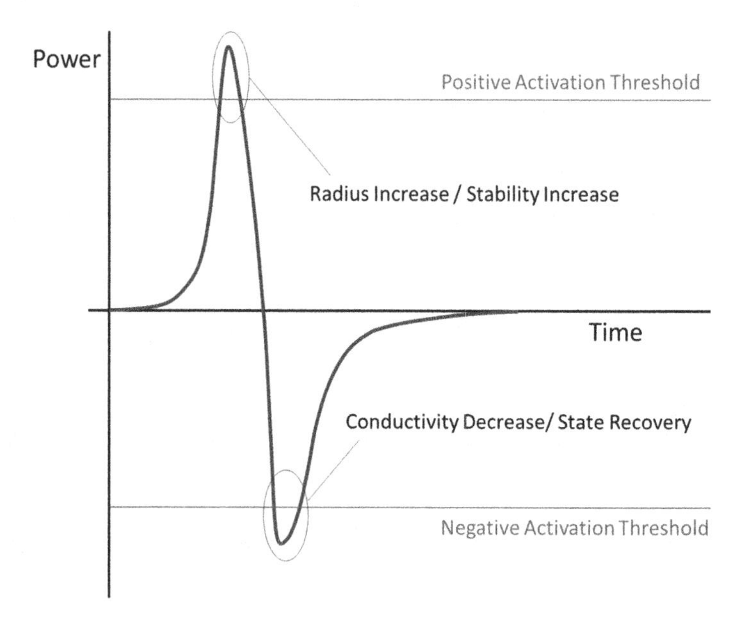

Fig. 9.5 The resistance state or synaptic weight can be made more stable with an electrical pulse that first increases the radius of the filament (positive power) then returns to the desired resistance state by decreasing conductivity (negative power)

can be created in artificial synapses in response to a positive outcome. Such a signal is illustrated in Fig. 9.5 in the form of an electrical pulse with positive polarity followed by a pulse with negative polarity. The positive portion of the pulse is required to increase the radius of the filament (and therefore stability) and the negative portion is required to raise the resistance back to its previous value.

The result is that, for electrical signals within the neural network having a range of powers, the state of the synapse can be moved from a plastic state to a more stable state through the use of controlled electrical signals. This addresses the stability-plasticity dilemma at the single device (synapse) level. Using this technique, neural networks can be designed that both adapt rapidly to unknown scenarios and maintain their expertise in addressing known ones.

9.3.5 Aging in Memristors

The above discussion suggests that any time a filament receives a positive polarity power greater than some threshold its radius increases and therefore its stability increases and plasticity decreases. This again is analogous to behavior in the biological systems. As we age so do our synapses. Youthful synapses are highly plastic and capable of learning rapidly whereas aged synapses are characterized by decreased rates of learning [46]. For many of us this feels like a regrettable side-effect of growing older but it is a very important component of neural development. Without this aging effect we would forget too rapidly and without youthful plasticity we would learn too slowly.

In memristor-based synapses this balance is mostly determined by the power of the applied electrical signals. A large power pulse can increase the radius of the filament substantially, even in a single event. Following that impulse, the synapse will appear to have aged significantly even though it has only experienced a single large pulse. If the amplitudes of the pulses are controlled by some probability distribution then the number of devices that have received a high amplitude pulse at some point in their history will follow a related distribution. The number of synapses that have received high amplitude pulses would then grow with time such that an ageing process similar to biological systems would occur in an aggregate sense.

An alternative approach exists where the stabilizing signals are controlled individually instead of by a probability distribution. The stabilizing pulses could be specified to increase the stability or radius of a given synapse by a small amount per stabilizing event. In that way, the number of stabilizing pulses received would determine the "age" of the synapse. Many other approaches to control aging can be envisioned but they all involve progressively increasing the stability of the synapse by increasing the radius at the expense of plasticity.

9.3.6 Synaptogenesis and Neurogenesis in Memristors

The knowledge that aging effects can be observed in memristor-based synapses is very encouraging. It suggests that neuromorphic neural networks can be designed that are capable of neural development similar to that in the brain and that the artificial network can progressively eliminate the process of forgetting as the things it learns become reinforced. It also presents potential challenges in that a deployed system may gradually lose its ability to respond to a new situation. In many cases this acquisition of expertise at the expense of the adaptive learning is not a problem, but in adaptable systems there needs to be some mechanism for recovering a youthful and highly plastic state.

As discussed in the previous section, aging in the brain is associated with a decrease in plasticity but clearly, as humans, we are capable of continued learning throughout our lives. There are many explanations for this continued ability but among the most interesting and highly researched is adult neurogenesis, where new neurons are generated to provide the needed plasticity in adulthood [47]. Many high-level functions have been correlated to adult neurogenesis and the dentate gyrus region in which it occurs [48, 49]. Chief among them is spatial mapping [50] where we frequently need to acquire and remember new information about our surroundings and environment. (Note: From preliminary experiments testing the ability to remember where our cars are parked, the authors suggest that this capability is still under-developed in humans but it should be possible to tailor the rate of aging in memristor-based neurons to achieve a desired rate of expertise.) Adult neurogenesis appears to present a unique and powerful neural computation algorithm for gradually specializing a neural structure to optimally represent inputs and contexts as they change over time [51, 52].

A remaining issue is the generation of new highly-plastic synapses or neurons. Perhaps the simplest method is to leave a bank of previously unused synapses that can be introduced to the network in the event that an unrecognized stimulus is encountered. Conceptually this is a simple approach but it presents challenges in implementation. For one, the network must be manufactured with a sufficient quantity of latent componentry and so may require a larger number of synapses and a larger network than other approaches. Second, this approach may require creative architectural solutions in order for the new neurons to have sufficient connectivity to existing neurons such that the new solutions can benefit from the existing expertise.

An alternative approach involves resetting a number of aged neurons to be replaced with new ones [53]. Some benefits of this approach are that the full network can be used at any one time to address the current range of problems and that newly generated synapses are natively incorporated within the previously acquired expertise of the aged network. Synapses to be killed can be either selected at random or chosen from among those regions that are activated by a specific set of stimuli.

Similar to the stabilizing pulses that lead to morphology change in conductive filaments, a specific pulse can be used to kill a synapse and cause neurogenesis. A large negative polarity pulse can be used to empty the oxygen vacancies, thereby putting the synapse in a very high resistance state. The device can be left in this state which can be both high resistance and stable, and such a state is a reasonable approximation for a non-existent synapse. Subsequent positive power pulses applied to this 'quasi-virgin' state act similarly to electroforming events, creating a small-radius/large-conductivity filament within the vacated region, thus restarting the evolution of memristive synaptic plasticity. An illustration of a pulse that would achieve such an effect is illustrated in Fig. 9.6.

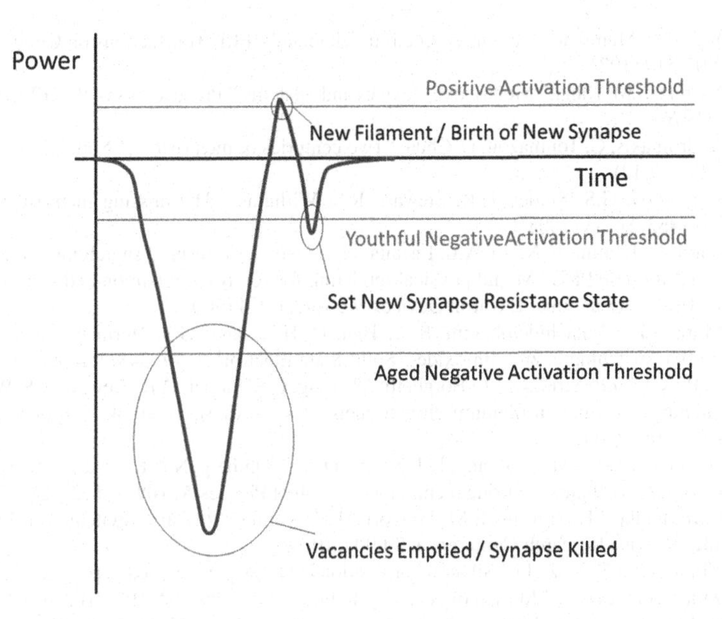

Fig. 9.6 An aged synapse can be emptied of vacancies or "killed" by a large power negative polarity electrical signal. When followed by a small positive polarity pulse this process can be used to turn a stable aged synapse into a youthful and highly plastic one

9.4 Conclusions

The memristor formalism provided by Leon Chua and promoted by Hewlett-Packard Labs has provided a compelling analogy to biological synapses and has led to very rapid progress in the field but it misses much of the complexity that is present in resistive switches. Examining this complexity, it is clear that these devices are in fact much more similar to biological synapses than was previously imagined and a variety of biomimetic opportunities exist for designing neural networks. By leveraging these advanced biomimetic functionalities, the use of memristors in neural networks (and other neuromorphic architectures) shows strong potential as an adaptive and accurate cyberthreat identification solution.

Sandia National Laboratories is a multi-program laboratory managed and operated by Sandia Corporation, a wholly owned subsidiary of Lockheed Martin Corporation, for the U.S. Department of Energy's National Nuclear Security Administration under contract DE-AC04-94AL85000.

References

1. W.S. Mcculloch, W. Pitts, "A logical calculus of the ideas immanent in nervous activity," Bulletin of Mathematical Biophysics, **5**, 115-133 (1943).
2. R. Kozma, R.E. Pino, G.E. Pazienza, "Advances in Neuromorphic Memristor Science and Applications," Springer (2012).

3. L.O. Chua, "Memristor – the missing circuit element," IEEE Transactions on Circuit Theory, **18**, 507-519 (1971).
4. L.O. Chua, S.M. Kang, "Memristive Devices and Systems," Proceedings of the IEEE, **64**, 209-223 (1976).
5. T. Prodromakis, C. Toumazou, L. Chua, "Two centuries of memristors," Nature Materials, **11**, 478-481 (2012).
6. D.B. Strukov, G.S. Snider, D.R. Stewart, R.S. Williams, "The missing memristor found," Nature, **453**, 80-83 (2008).
7. C. Cagli, D. Ielmini, F. Nardi, A.L. Lacaita, "Evidence for threshold switching in the set process of NiO-based RRAM and physical modeling for set, reset, retention and disturb prediction," IEEE International Electron Devices Meeting, p 1-4 (2008).
8. S. Murali, J.S. Rajachidambaram, S.-Y. Han, C.-H. Chang, G.S. Herman, J.F. Conley Jr, "Resistive switching in zinc-tin-oxide," Solid-State Electronics, **79**, 248-252 (2013).
9. M.D. Pickett, D.B. Strukov, J.L. Borghetti, J.J. Yang, G.S. Snider, D.R. Stewart, R.S. Williams, "Switching dynamics in titanium dioxide memristive devices," Journal of Applied Physics, **106**, 074508 (2009).
10. B.J. Choi, J.J. Yang, M.-X. Zhang, K.J. Norris, D.A.A. Ohlberg, N.P. Kobayashi, G. Medeiros-Ribeiro, R.S. Williams, "Nitride memristors," Applied Physics A, **109**, 1-4 (2012).
11. M.J. Marinella, J.E. Stevens, E.M. Longoria, P.G. Kotula, "Resistive switching in aluminum nitride," Device Research Conference, 89-90 (2012).
12. M. Mitkova, M.N. Kozicki, "Silver incorporation in Ge-Se glasses used in programmable metallization cell devices," Journal of Non-Crystalline Solids, **299-302**, 1023-1027 (2002).
13. M.N. Kozicki, M. Balakrishnan, C. Gopalan, C. Ratnakumar, M. Mitkova, "Programmable Metallization Cell Memory Based on Ag-Ge-S and Cu-Ge-S Solid Electrolytes," Non-Volatile Memory Technology Symposium, 83-89 (2005).
14. R. Waser, M. Aono, "Nanoionics-based resistive switching memories," Nature Materials, **6**, 833-840 (2007).
15. L.O. Chua, "Resistance switching memories are memristors," Applied Physics A, **102**, 765-783 (2011).
16. J.J. Yang, D.B. Strukov, D.R. Stewart, "Memristive devices for computing," Nature Nanotechnology, **8**, 13-24 (2013).
17. H.S.P. Wong, H.-Y. Lee, S. Yu, Y.-S. Chen, Y. Wu, P.-S. Chen, B. Lee, F.T. Chen, M.-J. Tsai, "Metal-Oxide RRAM," Proceedings of the IEEE, **100**, 1951-1970 (2012).
18. J. Hutchby, M. Garner, "Assessment of the Potential & Maturity of Selected Emerging Research Memory Technologies Workshop & ERD/ERM Working Group Meeting," (2010).
19. P.R. Mickel, A.J. Lohn, B.J. Choi, J.J. Yang, M.-X. Zhang, M.J. Marinella, C.D. James, R.S. Williams, "A physical model of switching dynamics in tantalum oxide memristive devices," Applied Physics Letters, **102**, 223502 (2013).
20. J.P. Strachan, A.C. Torrezan, F. Miao, M.D. Pickett, J.J. Yang, W. Yi, G. Medeiros-Ribeiro, R.S. Williams, "State Dynamics and Modeling of Tantalum Oxide Memristors," IEEE Transactions on Electron Devices, **60**, 2194-2202 (2013).
21. A.J. Lohn, P.R. Mickel, M.J. Marinella, "Dynamics of percolative breakdown mechanism in tantalum oxide resistive switching," Applied Physics Letters, **103**, 173503 (2013).
22. H.Y. Lee, Y.S. Chen, P.S. Chen, T.Y. Wu, F. Chen, C.C. Wang, P.J. Tzeng, M.-J. Tsai, C. Lien, "Low-Power and Nanosecond Switching in Robust Hafnium Oxide Resistive Memory With a Thin Ti Cap, IEEE Electron Device Letters, **31**, 44-46 (2010).
23. S. Lee, W.-G. Kim, S.-W. Rhee, K. Yong, "Resistance Switching Behaviors of Hafnium Oxide Films Grown by MOCVD for Nonvolatile Memory Applications," Journal of the Electrochemical Society, **155**, H92-H96 (2008).
24. D.O. Hebb, "The Organization of Behvior: A Neuropsychological Theory," Wiley, (1949).
25. I.E. Ebong, P. Mazumder, "CMOS and Memristor-Based Neural Network Design for Position Detection," Proceedings of the IEEE, **100**, 2050-2060 (2012).
26. A. Thomas, "Memristor-based neural networks," Journal of Physics D: Applied Physics, **46**, 093001 (2013).

27. N. Brunel, V. Hakim, "Fast Global Oscillations in Networks of Integrate-and-Fire Neurons with low Firing Rates," Neural Computation, **11**, 1621-1671 (1999).
28. A.L. Hodgkin, A.F. Huxley, "A quantitative description of membrane current and its application to conduction and excitation in nerve," The Journal of Physiology **117** (4) 500-544 (1952).
29. M.D. Pickett, G. Medeiros-Ribeiro, R.S. Williams, "A scalable neuristor built with Mott memristors," Nature Materials, **12**, 114-117 (2012).
30. C. Clopath, W. Gerstner, "Voltage and spike timing interact in STDP – a unified model," Frontiers in Synaptic Neuroscience, **2**, 1-11 (2010).
31. N. Caporale, Y. Dan, "Spike Timing-Dependent Plasticity: A Hebbian Learning Rule," Annual Review of Neuroscience, **31**, 25-46 (2008).
32. D.E. Feldman, "The Spike-Timing Dependence of Plasticity," Neuron, **75**, 556-571 (2012).
33. S. Ambrogio, S. Balatti, F. Nardi, S. Facchinetti, D. Ielmini, "Spike-timing dependent plasticity in a transistor-selected resistive switching memory," Nanotechnology, **24**, 384012 (2013).
34. F. Miao, J.P. Strachan, J.J. Yang, M.-X. Zhang, I. Goldfarb, A.C. Torrezan, P. Eschbach, R.D. Kelley, G. Medeiros-Ribeiro, R.S. Williams, "Anatomy of a Nanoscale Conduction Channel Reveals the Mechanism of a High-Performance Memristor," Advanced Materials, **23**, 5633-5640 (2011).
35. G.-S. Park, Y.B. Kim, S.Y. Park, X.S. Li, S. heo, M.-J. Lee, M. Chang, J.H. Kwon, M. Kim, U.-I. Chung, R. Dittmann, R. Waser, K. Kim, "In situ observation of filamentary conducting channels in an asymmetric Ta2O5-x/TaO2-x bilayer structure," Nature Communications, **4**, 1-9 (2013).
36. D.B. Strukov, F. Alibart, R.S. Williams, "Thermophoresis/diffusion as a plausible mechanism for unipolar resistive switching in metal-oxide-metal memristors," Applied Physics A, **107**, 509-518 (2012).
37. F. Miao, W. Yi, I. Goldfarb, J.J. Yang, M.-X. Zhang, M.D. Pickett, J.P. Strachan, G. Medeiros-Ribeiro, R.S. Williams, "Continuous Electrical Tuning of the Chemical Composition of TaOx-Based Memristors," ACS Nano, **6**, 2312-2318 (2012).
38. P.R. Mickel, A.J. Lohn, M.J. Marinella, "Isothermal Switching and Detailed Filament Evolution in Memristive Systems," Advanced Materials, **26**, 4486-4490 (2014).
39. A.J. Lohn, P.R. Mickel, C.D. James, M.J. Marinella, "Degenerate Resistive Switching and Ultrahigh Density Storage in Resistive Memory," Applied Physics Letters, **105**, 103501 (2014).
40. P.R. Mickel, A.J. Lohn, M.J. Marinella, "Precise electrical control of nanoscale resistive filament geometry," unpublished (2013).
41. D.F. Marrone, T.L. Petit, "The role of synaptic morphology in neural plasticity: structural interactions underlying synaptic power," Brain Research Reviews, **38**, 291-308 (2002).
42. R. Yuste, T. Bonhoeffer, "Morphological Changes in Dendritic Spines Associated with Long-Term Synaptic Plasticity," Annual Reviews of Neuroscience, **24**, 1071-1089 (2001).
43. R. Lamprecht, J. LeDoux, "Structural Plasticity and Memory," Nature Reviews Neuroscience, **5**, 45-54 (2004).
44. H. Kasai, M. Fukuda, S. Watanabe, A. Hayashi-Takagi, J. Noguchi, "Structural dynamics of dendritic spines in memory and cognition," Trends in Neurosciences, **33**, 121-129 (2010).
45. M. Matamales, "Neuronal activity-regulated gene transcription: how are distant synaptic signals conveyed to the nucleus?," F1000Research, **1**, 69 (2012).
46. S.N. Burke, C.A. Barnes, "Neural plasticity in the ageing brain," Nature Reviews Neuroscience, **7**, 30-40 (2006).
47. F.H. Gage, "Neurogenesis in the Adult Brain," The Journal of Neuroscience, **22**, 612-613 (2002).
48. J.B. Aimone, J. Wiles, F.H. Gage, "Potential role for adult neurogenesis in the encoding of time in new memories," Nature Neuroscience, **9**, 723-727 (2006).
49. I. Imayoshi, M. Sakamoto, T. Ohtsuka, K. Takao, T. Miyakawa, M. Yamaguchi, K. Mori, T. Ikeda, S. Itohara, R. Kageyama, "Roles of continuous neurogenesis in the structural and functional integrity of the adult forebrain," Nature Neuroscience, **11**, 1153-1161 (2008).
50. C.D. Clelland, M. Choi, C. Romberg, G.D. Clemenson Jr, A. Fragniere, P. Tyers, S. Jessberger, L.M. Saksida, R.A. Barker, F.H. Gage, T.J. Bussey, "A Functional Role for Adult Hippocampal Neurogenesis in Spatial Pattern Separation," Science, **325**, 210-213 (2009).

51. J.B. Aimone, J. Wiles, F.H. Gage, "Computational Influence of Adult Neurogenesis on Memory Encoding," Neuron, **61**, 187-202 (2009).
52. Y. Li, J.B. Aimone, X. Xu, E.M. Callaway, F.H. Gage, "Development of GABAergic inputs controls the contribution of maturing neurons to the adult hippocampal network," Proceedings of the National Academy of Science, **109**, 4290-4295 (2012).
53. R.A. Chambers, M.N. Potenza, R.E. Hoffman, W. Miranker, "Simulated Apoptosis/ Neurogenesis Regulates Learning and Memory Capabilities of Adaptive Neural Networks," Neuropsychopharmacology, **29**, 747-758 (2004).

Chapter 10
Low Power Neuromorphic Architectures to Enable Pervasive Deployment of Intrusion Detection Systems

Tarek M. Taha, Raqibul Hasan, Chris Yakopcic, and Mark R. McLean

10.1 Introduction

Intrusion detection systems (IDS) are commonly utilized to prevent cyber-attacks. With the wide proliferation of network connected devices, running IDS algorithms on all devices (including mobile devices) within a network can help bolster security. However, the cost of running IDS algorithms on all networked devices can be high in terms of power and physical resources (especially battery operated ones). Several recent studies have proposed mapping applications to neural network form and then running these on specialized neural network accelerators [1, 2] to reduce power consumption. Neural accelerators can result in power reduction from about two times to several thousand times compared to RISC processors [3]. Hence utilizing these neural network accelerators can enable the deployment of IDS algorithms across all devices in a network.

Neural networks are very efficient for pattern recognition tasks. Given that the intrusion detection problem is fundamentally a classification problem, several studies have proposed solving IDS algorithms using neural networks [4–6]. Bass [4] presented a RBF neural network based IDS. Jiang and Yu [6] presented two supervised methods, the support vector machine as well as multi-layered neural networks, for IDS. Their results show a high detection rate and low false alarm rate.

This chapter examines the design of novel specialized multicore neural processors, where the processing cores would be similar to the architecture in Fig. 10.1. We studied the impact of using two memory technologies for the synaptic array in Fig. 10.1: SRAM and memristors. Memristors [7–9] are a new class of devices that

T.M. Taha (✉) • R. Hasan • C. Yakopcic
University of Dayton, Dayton, OH 45469, USA
e-mail: tarek.taha@udayton.edu; hasanm1@udayton.edu; cyakopcic1@udayton.edu

M.R. McLean
Center for Exceptional Computing, Baltimore, MD 21202, USA
e-mail: mrmclea@lps.umd.edu

R.E. Pino et al. (eds.), *Cybersecurity Systems for Human Cognition Augmentation*, Advances in Information Security 61,
DOI 10.1007/978-3-319-10374-7_10

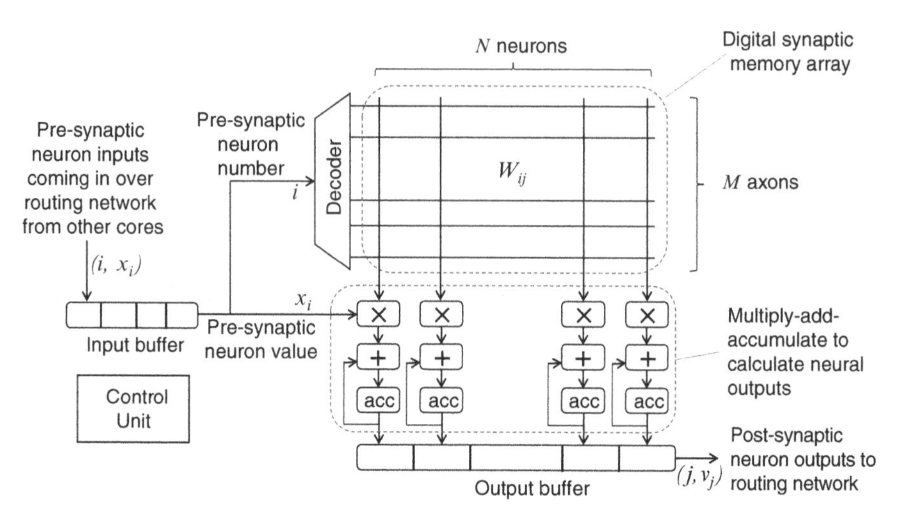

Fig. 10.1 Proposed neural core architecture

can offer lower energy and higher throughput per chip. Two types of memristor cores were examined: digital and analog cores. Novel circuits were designed for both of these memristor systems. Very detailed circuit simulations were used to ensure that the systems could be compared accurately. The memristor circuits were simulated using an accurate memristor SPICE model we developed recently [10].

The impact of scaling to a multicore chip (as shown in Fig. 10.2) was studied for each form of the core in Fig. 10.1. To ensure accurate modeling of the multicore system, we designed specialized routing circuits between the cores and accounted for their area, power, and delays. Our results show that the top two systems with highest throughput and lowest power are the analog and then digital memristor systems. The SRAM based digital system is third best. It has higher energy cost due to leakage energy—the memristors do not have this, thus making them more energy efficient.

We compared these specialized systems to more traditional HPC systems. Two commodity high performance processors were examined: a six core Intel Xeon processor, and an NVIDIA Tesla M2070 GPGPU. Care was taken to ensure the code on each platform was very efficient: multi-threaded on the Xeon processor to utilize all six cores, and a highly parallel CUDA program on the GPGPU. Our results indicate that the specialized systems can be between two to five orders more energy efficient compared to the traditional HPC systems. The energy efficiency depends on the options utilized within the core (such as SRAM or memristor memory and data bit width). The specialized cores take up much less die area—allowing, in some cases, a reduction from 179 Xeon six-core processor chips down to 1 memristor based multicore chip and a corresponding reduction in power from 17 kW down to 0.07 W.

In this study we examine the properties of several neural cores for implementing a multi-layered neural network. Each neuron in the neural network performed two types of operations: it summed its weighted inputs and evaluated a non-linear function. If the axonal inputs to a neuron are given by x_i then the corresponding neuron output was evaluated as:

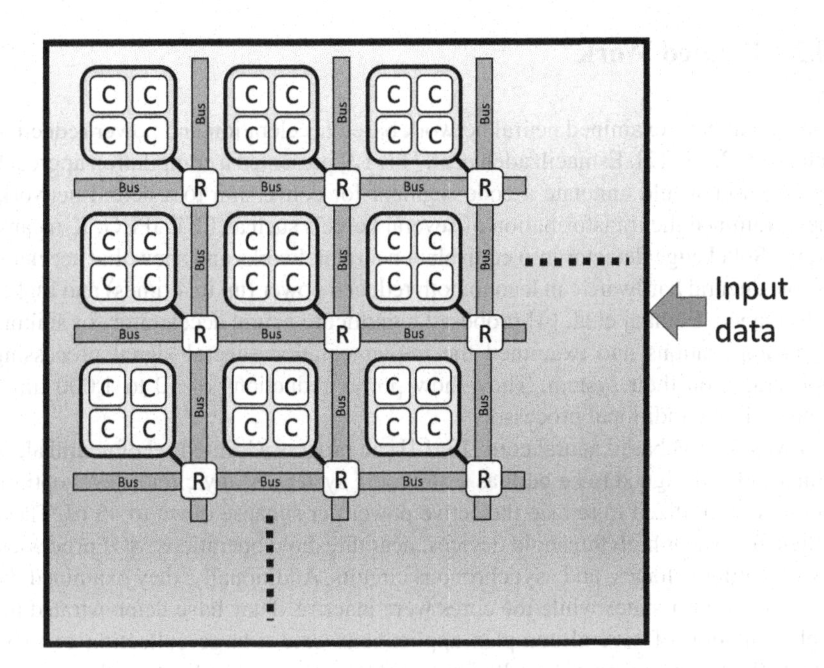

Fig. 10.2 Proposed multicore system with several neural cores (C) sharing a routing resource (R)

$$v_j = \sum_i W_{i,j} x_i \tag{10.1}$$

$$y_j = f(v_j) \tag{10.2}$$

Here, W is a weight matrix in which $W_{i,j}$ is the synaptic weight of axon i for neuron j and f is a nonlinear function (usually a sigmoid function).

The novel contributions of this work are:

1) A detailed design of two memristor neural cores: an analog and a digital core. The timing and power data for both cores are based on detailed SPICE simulations using a novel SPICE model we developed [10]. To the best of our knowledge, this is the first detailed design and SPICE simulation of a memristor core in the literature.
2) Novel circuit designs were developed for the memristor cores to reduce their overall power consumption and enable fast processing. These techniques cut down the power consumption at the expense of an increase in the area of the core.
3) The area, power, and performance of a full multicore chip based on our design were estimated through detailed modeling and simulations. Area and power of an efficient on-chip routing system was examined to obtain a more accurate evaluation of the full chip.
4) A comparison of the specialized cores against current high end processing platforms was carried out. This gives an idea of the power and area savings that can be achieved through the use of specialized cores. A comparison of this form for memristive systems has not been carried out before.

10.2 Related Work

Studies that have examined neural network based acceleration and power reduction include [1, 2, 11, 12]. Esmaeilzadeh et al. [11, 12] presented a compilation approach where a user would annotate a code segment for conversion to a neural network. They examined the transformation of several kernels such as FFT, JPEG, K-means, and the Sobel edge detector into equivalent neural networks and show that acceleration on a neural hardware can lead to both reduced power (up to 4 times) and higher performance. Belhadj et al. [4] proposed a multicore neural accelerator for spiking neural applications and examined the implementation several signal processing applications on their system. They show power reductions of 10 to 1,000 times compared to a traditional processor.

IBM's SRAM based neural core [13, 14] can process a limited set of neural algorithms and is designed to be built as a multicore system. Many circuit-level optimizations were utilized to reduce the active power per synapse down to 45 pJ. These include the use of high threshold devices, near threshold operations, SOI processes, low operating voltages, and asynchronous circuits. Additionally, they examined the use of low power states while the cores were inactive. They have demonstrated the implementation of several complex applications onto a large collection of these cores [2]. As an example, a collision avoidance system application they implemented utilized over 21,000 hardware neural cores.

10.3 Architectural Overview

Figure 10.1 shows a block diagram of our proposed digital core for processing the feed forward neural network described in Eq. 10.1. Each core processes a collection of N neurons, with each neuron having up to M input axons. The input synaptic weights ($W_{i,j}$) are stored in a memory array. These synaptic values are multiplied with the presynaptic input values (x_i) and are summed into an accumulator. Once the final output neural values are generated, they are sent to other cores for processing. Input and output buffers store the presynaptic inputs and post synaptic outputs respectively.

We assumed a multicore architecture as shown in Fig. 10.2. A collection of neural processing cores are grouped together and connected to other groups of processing cores through an on-chip routing network. A large neural network would be distributed across multiple cores, with each core processing at most N neurons from the network.

The memory array shown in Fig. 10.1 can be developed using several different memory technologies. The most common memory technology integrated with processing cores today is SRAM. In SRAM technology, each memory element utilizes six transistors and stores a binary value (either 0 or 1). Resistive memory technologies, such as STT-MRAM, PCRAM, and memristors, are increasingly being studied as replacements for SRAM. This is primarily because of their higher densities and nonvolatility. Of these technologies, memristors have the highest range of resistance values and have the potential for the highest memory density. The high resistance range in memristor devices allows them to model synapses in analog form. In this form, multiply and add operations can be carried out using memristor resistance values and current summation respectively, thus reducing circuit area and power consumption.

In this study we examine the design of the neural core in Fig. 10.1 with both SRAM and with memristor devices. Detailed analysis is used to determine the area, power, and timing of each core [16]. Two classes of memristor based neural cores are examined: a digital case identical to the design in Fig. 10.1, and an analog case that was a slight variant of the design in Fig. 10.1. For all three cases (SRAM, memristor digital, and memristor analog), we examine system properties with 1 bit per neuron and with 4 bits per neuron. For the 1 bit per neuron systems, we assumed that the synapses are 2 bits wide. For the 4 bits per neuron systems, we assumed that the synapses are 4 bits wide.

10.4 SRAM Neural Core

The SRAM based neural core studied is identical to the architecture in Fig. 10.1 (somewhat similar to the IBM study [13, 14]). We assumed that each core would process 256 neurons and have 1024 axons for each neuron. This would require the memory array to hold the data for 1024×256 synapses. Two specific cases of the core were examined: one with 2 bit synapses and one with 4 bit synapses. The 2 bit synapse system had a 1024×512 bit SRAM array and used 12 bit adders to accumulate weights. The 4 bit synapse system had a 1024×1024 bit SRAM array and used 4 bit multipliers and 18 bit adders.

We assumed all systems were to be developed using a 45 nm process for this study and operated at a 200 MHz clock frequency. The area, power, and timing of the SRAM array was calculated using the CACTI cache modeling tool [15] with the low operating power transistor option utilized. Components of a typical cache that would not be needed in the neural core were not included in the area and power calculations. These include the tag array and tag comparator power and area. The power and timing of the multipliers, adders, and all registers were calculated using SPICE. The area, power, and timing of the routing core were calculated using the Orion [16] interconnection network tool (assuming 8 bits per link). The area, power, and timing of the input and output buffers were calculated based on the buffer calculations done in Orion. The control logic within the neural core was assumed to be equivalent to the router control logic area and power (as calculated by Orion). Although this area was simply an approximation, we believe this does not affect the results of this study significantly as the SRAM array dominates the area and power of the core.

10.5 Memristor Neural Core

Two types of memristor based neural cores were examined: a digital type identical to the design in Fig. 10.1, and an analog type with some design optimizations compared to Fig. 10.1. A crossbar structure was utilized for the memristor memory circuits in this study. We developed several novel approaches for designing the memristor cores. Detailed SPICE level simulations of these circuits were utilized to examine their power and timing properties.

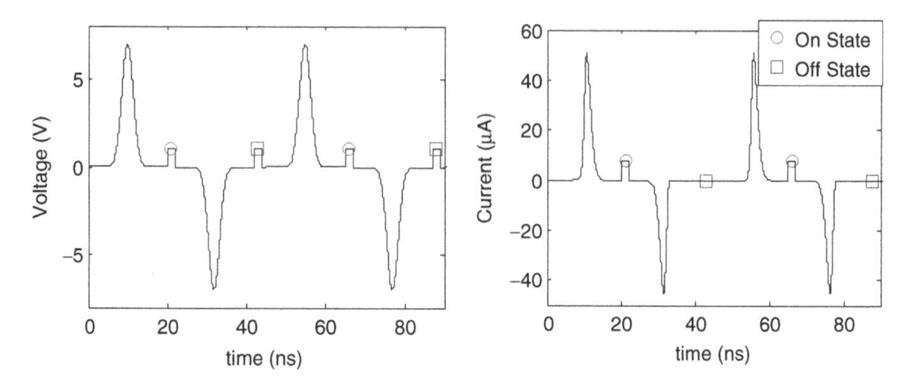

Fig. 10.3 Simulation results displaying the input voltage and current waveforms. The following parameter values were used in the model [15] to obtain this result: Vp = 1.088 V, Vn = 1.088 V, Ap = 816,000, An = 816,000, xp = 0.985, xn = 0.985, αp = 0.1, αn = 0.1, a1 = 1.6(10-4), a2 = 1.6(10-4), b = 0.05, x0 = 0.01

10.5.1 Memristor Model

To perform a device level analysis of a memristor crossbar memory system, a SPICE equivalent of the memristor model first proposed in [10] was utilized. Previous work [10] shows that this model correlates well with many published devices. This model was set to match the characterization data of one of the memristor devices published in [17] (see Fig. 10.3). This device was chosen because it had a large R_{OFF}/R_{ON} ratio (10^6) while still retaining a relatively low switching time (about 10 ns). It also has a large on state resistance of about 125kΩ (determined by the 8 μA current from a 1 V read pulse).

The simulation result in Fig. 10.3 shows the minimum and maximum resistances of the model to be 124.95kΩ and $125.79 \times 10^9 \Omega$ respectively, which correlates very closely to the characterization [17]. These strong simulation results show that a reliable device model has been developed, and this will lead to more accurate results when simulating the crossbar tiles.

10.5.2 Memristor Crossbar Digital Core

Table 10.1 lists the read energy consumption of different sized memristor crossbar arrays. These power values were generated through detailed SPICE simulations taking wire resistance into account. It is seen that as the crossbar size grows, the read energy increases. This is due to extra sneak paths in a larger memory array. To constrain the read energy of a large memristor memory, we developed a tiled crossbar memory architecture as shown in Fig. 10.4. In this case, a single transistor is placed on each input and each output of the crossbar to allow a tile to be isolated from other

Table 10.1 Read and write energy of memristor crossbars of different sizes

	Training energy per synapse (pJ)	Read energy per synapse (fJ)
4×4 Crossbar	1.90	1.84
8×8 Crossbar	5.07	2.00
Untiled	100.3	21

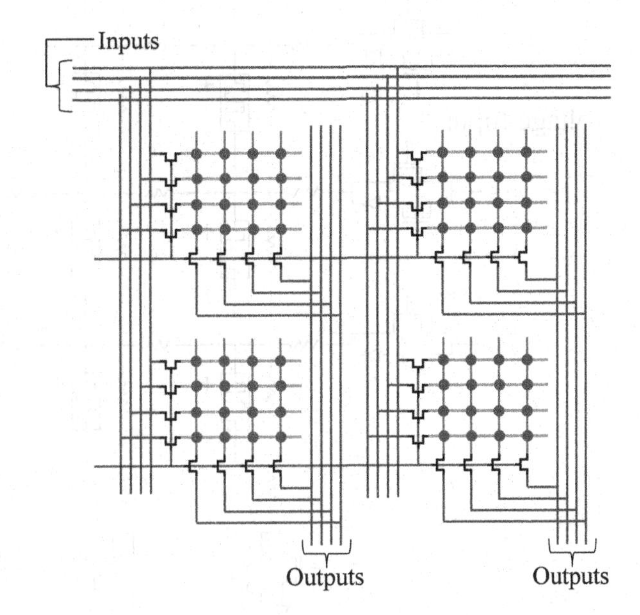

Fig. 10.4 Tiled memristor crossbar array

tiles. Only one row of tiles is accessed at a time and this limits the leakage current within the memristor grid. The leakage current present during an operation is now limited to that of a 4×4 crossbar. This tiling approach reduces the number of leakage paths and thus reduces the overall dynamic power consumption of the memory compared to an untiled crossbar array (see Table 10.1). In the digital memristor memory based neural core, the SRAM array was replaced with the tiled crossbar array seen in Fig. 10.4. All other components were assumed to remain the same.

10.5.3 Neural Computations in an Analog Crossbar Array

Given that memristors have a large range of resistance values (R_{OFF}/R_{ON} ratio of 10^6 for the device we simulated), it is possible for a single memristor to encode a large range of synaptic weights in analog form. Thus a crossbar of memristors could potentially simulate a neural network inherently. To demonstrate this, we trained a 4×4 crossbar array (see Fig. 10.5) to classify a set of 4×4 binary patterns. Each row of a pattern was converted to an analog voltage and supplied as input to the crossbar. A detailed SPICE simulation of the crossbar that considered wire resistance showed that the circuit was able to correctly classify the linearly separable set of patterns

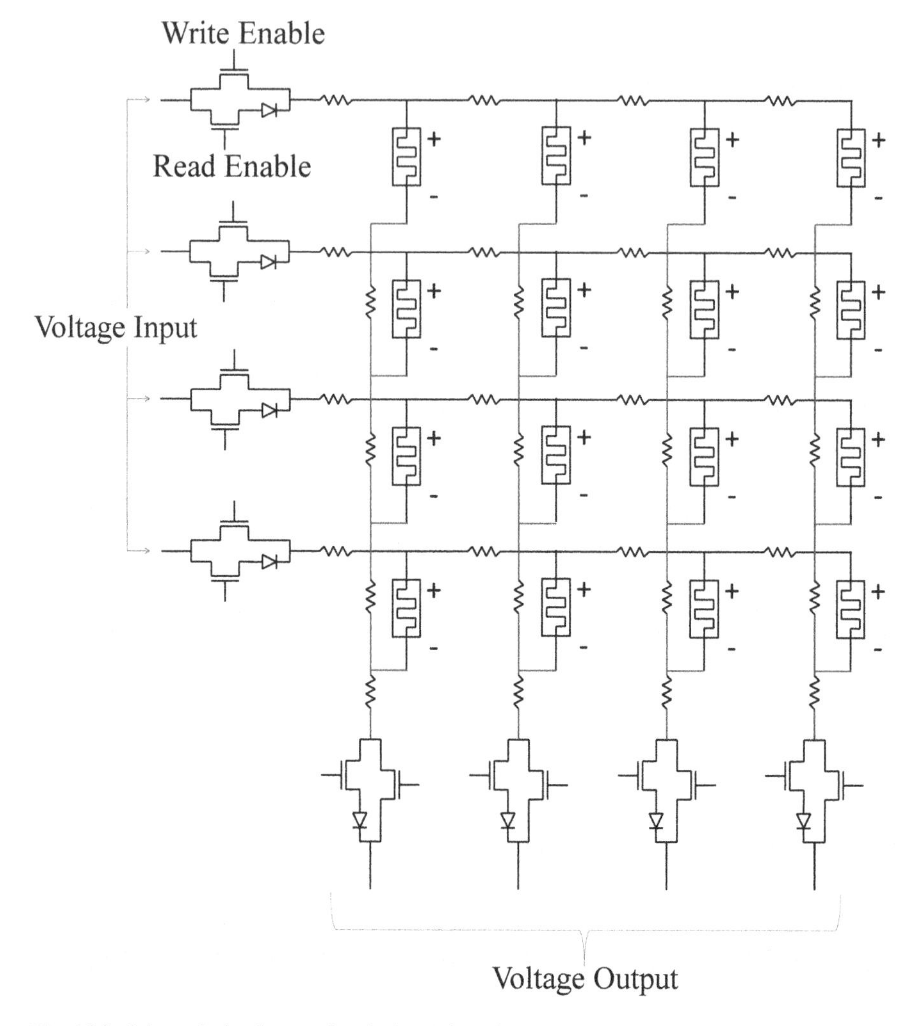

Fig. 10.5 Schematic for the crossbar designed for a 2 layer neural network simulation

(see Fig. 10.6). Each row in each of the input images is represented by a 16 value input pulse that is applied to one of the 4 inputs in the circuit in Fig. 10.5. The 4×4 crossbar is inherently carrying out multiply operations by the flow of current through a memristor resistance, and doing add operations by the summing of currents in an output column. The conductivity of the each of the memristors was trained iteratively so correct classification could be performed. This classification experiment is explained in more detail in [18]. At each input and output of the crossbar, an isolator circuit consisting of two transistors and a diode is placed as shown in Fig. 10.5. This enables a large array of these crossbars to be read simultaneously. This is discussed further towards the end of the next section.

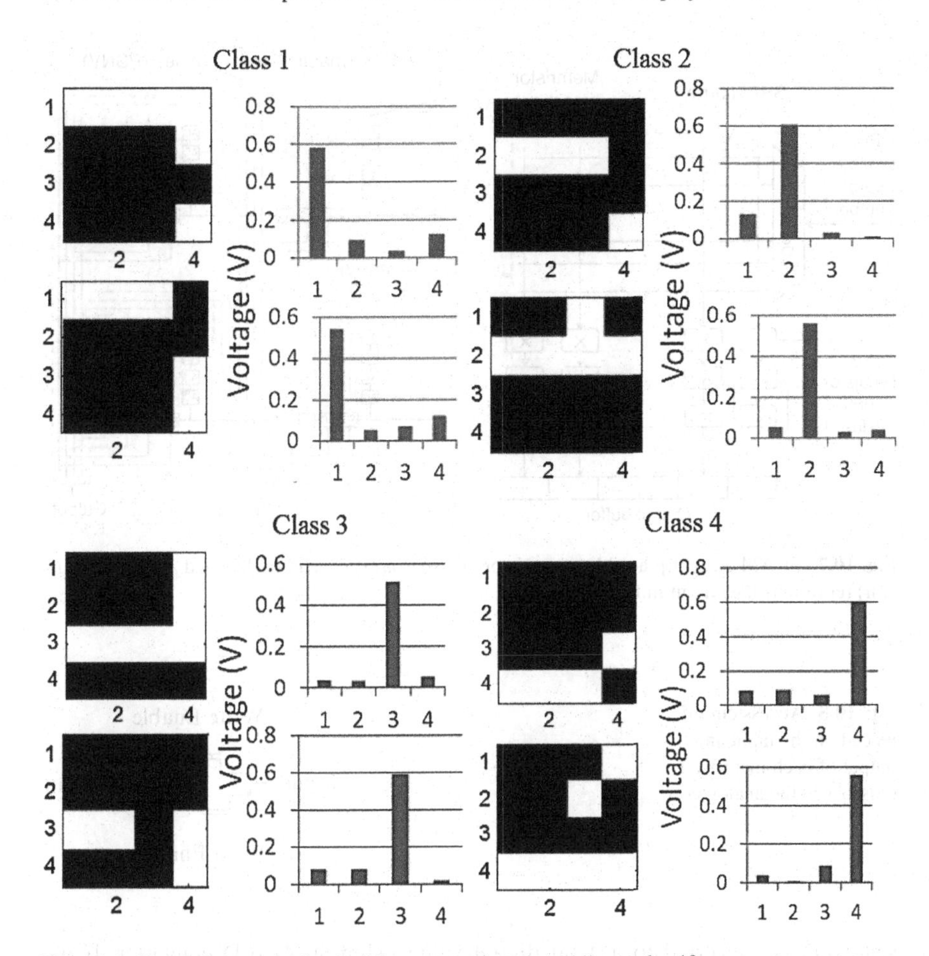

Fig. 10.6 Simulation results for the 2 layer memristor based linear classifier

10.5.4 Analog Memristor Neural Core

The analog memristor core also utilized a tiled memristor crossbar array as shown in Fig. 10.7. A tiled design was used as this would reduce the number of sneak paths and thus the overall energy consumption. This tiled configuration was designed to allow all tiles to be read simultaneously. Thus several components in the digital circuit could be eliminated, including the memory array decoder and the multiply-add-accumulate circuits. This parallel read operation allows the large crossbar to be much faster than the digital circuit.

As with the digital case, we consider systems with 1 bit and 4 bit neural values. In the system with 4 bit values, a D-to-A converter is needed for each input axonal value. At the output of the crossbar array, the current summed from all tiled crossbar

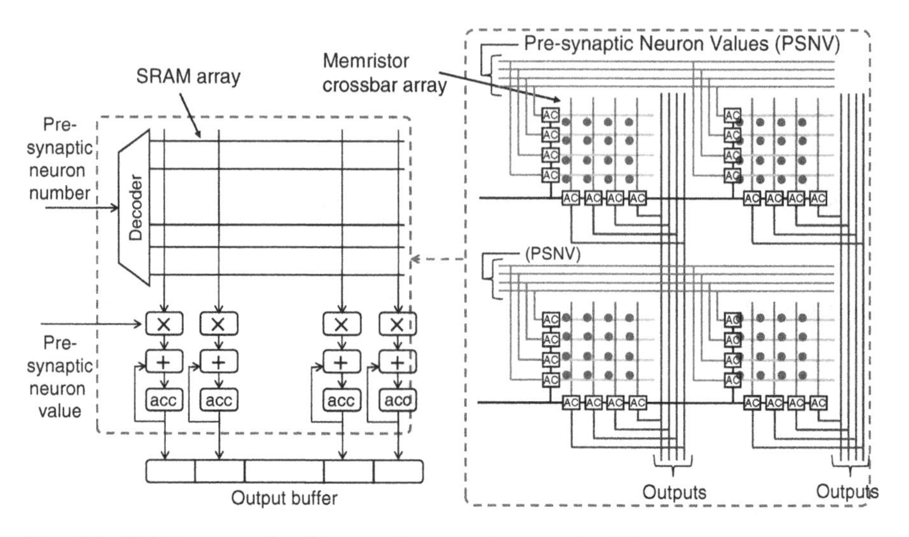

Fig. 10.7 SRAM array replaced by memristor crossbar array. Each box labelled AC (access circuit) represents the circuit in Fig. 10.8

Fig. 10.8 Access circuit placed on the input and output of each memristor crossbar in the analog core

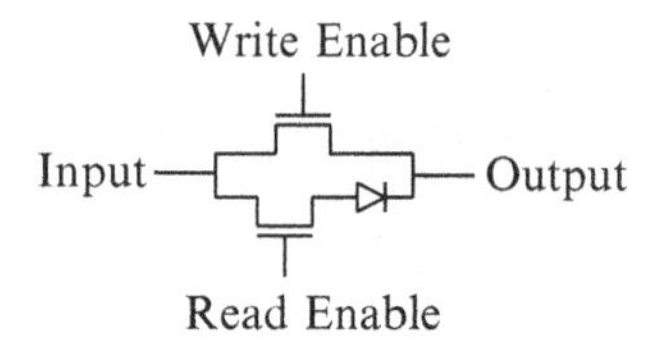

segments is converted to a 4 bit digital value using an A-to-D converter. If we assume each synapse can only hold a limited range of values, a small A-to-D circuit would be needed. In this study we optimized the design of the A-to-D and D-to-A circuits and examined their function, power, and timing using SPICE. Since it was decided that a maximum of 4 bits of precision was needed for the D-to-A and A-to-D circuits, their designs were simplified. They are each based on a set of 16 transistors that switch according to the analog input voltage level.

To avoid the large energy requirement and noise associated with a high density memristor crossbar [19], the memristor crossbar used in these simulation had been partitioned into smaller 8×8 tiles. This limits the energy consumption and the unwanted current paths in the system. To ensure that unwanted current paths do not flow between crossbars, the isolator circuit in Fig. 10.8 is used at each input and output of the crossbars. When the write switch is enabled, current can flow in either direction. During the training phase, these transistors are used to only turn on one row of 8×8 crossbars at a time to ensure no unwanted current paths are present. Within a row of crossbars, a training operation is applied to a single row of a crossbar similarly to the digital system.

When the read mode is enabled, current can only flow in one direction since a read pulse will always be positive. The diodes stop current from flowing incorrectly to a different 8 × 8 crossbar, so the entire memristor grid can be read in parallel using this approach. Alternate current paths are limited to the circuit within a tile and cannot occur between tiles.

10.6 Simulation on Commodity Processors

To assess the performance of the neural cores designed, we compared their performance against current commodity high performance processors. Two processor platforms were examined: an Intel Xeon X5650 processor and an NVIDIA Tesla M2070 GPGPU. The Xeon processor had six cores, ran at 2.66 GHz, and consumed 95 W maximum power. The GPGPU had 448 cores, ran at 575 MHz and consumed 225 W maximum power. We developed multi-layered feed-forward neural network codes to be run on these processors with care taken to ensure the highest throughput was obtained from each processor.

On the Xeon processor, a two layer network with 1,024 neurons/layer was simulated for 10 iterations. Each output neuron had 1,024 synapses. This network configuration was chosen as it would fit fully within the on-chip 12 MB level 3 cache and thus performance was not hampered by off-chip data access latency. This gives a more reasonable comparison with our neural cores where all synaptic data is stored on-chip as well. The neural network program was written in C and utilized the POSIX thread library to enable multi-threading. Thus all six cores on the chip were utilized. A throughput of 3.6 million neurons/s was achieved. Although our C program did not utilize the SIMD capability of the processors, we multiplied the neurons/s throughput by four to mimic the best possible performance that could have been achieved with the Intel Xeon processor.

GPGPUs provide significantly better performances compared to traditional multicore processors for data parallel applications. Several groups have studied the simulation of neural networks on these systems [20, 21]. We examined the simulation of a 10 layer neural network with 1,024 neurons/layer for 20 iterations on an NVIDIA Tesla M2070 GPGPU. The program was written in CUDA. It was parallelized to enable each thread to model one synapse, allowing a maximum of 1,024 × 1,024 threads to exist at a time. This high parallelism is needed to take advantage of the large number of cores on the GPGPU (448 cores). The program was designed to use shared memory to optimize performance. Profiling of the code using the CUDA profiler showed 85 % utilization of the GPGPU cores. The throughput of the GPGPU was about 215 million neurons per second. This is a 60 times speedup over the Xeon processor. However the higher power of the GPGPU makes the GPGPU about 6 times more power efficient per neuron processed compared to the Intel Xeon processor (assuming SIMD operations on the Xeon processor).

10.7 Results

10.7.1 Core Properties

We developed detailed area, timing, and power analysis of the neural cores studied. Two versions of each core were studied with neuron values being either 1 bit or 4 bit. Tables 10.2 and 10.3 show the three configurations examined for the 1 bit and 4 bit cases respectively. The energy per neuron listed in the tables is based on a comprehensive analysis of each core. They are based on both the active and leakage energy for all the components within each core and also the energy needed to transfer data into and out of the cores.

The 1 bit and 4 bit per neuron cases were designed to use 2 bits and 4 bits per synapse respectively. Thus the digital systems needed multiple memory cells to store a synaptic value. In the analog cores, it was assumed that the resistance of one memristor could be varied to accommodate these synaptic values. With 1 bit per neuron, a multiplier is not needed to calculate neural outputs. With 4 bits per neuron, multipliers are needed in the digital cores, while the analog cores use the memristor conductance to carry out multiplications in the analog domain. In all the digital cores, one synapse is read per cycle for all the neurons in parallel, thus requiring an adder to accumulate the weighted synaptic outputs over multiple cycles. In the analog cores, all the synapses of all the neurons operate in parallel, thus allowing current summations to perform addition of the synaptic outputs.

Table 10.2 One bit per neuron core configurations and performance (multiplier is not needed because of 1 bit input)

Configuration	Synaptic memory device	Memory cells per synapse	Bits per synapse	Adder	Core area (mm^2)	Energy per neuron (nJ)	Core throughput (neurons/s)
1	Memristor	1	Analog	Current add	0.037	0.02	263.9×10^6
2	Memristor	2	2 bits	12 bit adder	0.098	0.24	42.1×10^6
3	SRAM	2	2 bits	12 bit adder	0.288	0.38	42.1×10^6

Table 10.3 Four bit per neuron core configurations and performance

Configuration	Synaptic memory device	Memory cells per synapse	Bits per synapse	Multiplier	Adder	Core area (mm^2)	Energy per neuron (nJ)	Core throughput (neurons/s)
4	Memristor	1	Analog	Memr. resist.	Current add	0.058	0.03	66.3×10^6
5	Memristor	4	4 bits	4 bit multiplier	18 bit adder	0.179	0.39	28.6×10^6
6	SRAM	4	4 bits	4 bit multiplier	18 bit adder	0.513	0.78	26.6×10^6

10.7.2 System Properties

This section examines the system level impact of the core architectures described in Tables 10.2 and 10.3. We calculated the performance of the multicore systems based on the cores described earlier. Each system was assumed to have a routing network to allow multiple cores to communicate with each other. We assumed that eight cores shared a routing core.

Two examples of neural network processing scenarios were studied. In the first example, all systems processed a network with 25,600 neurons at a rate of 150,000 iterations per second. This leads to a throughput of 2.56 billion neurons processed per second. In the second example, we kept the neurons per second throughput the same as the first example, but processed a much larger network at a slower rate of 1,500 iterations/s. In addition, we examined what the runtime would be on a cluster of current high performance computer systems. The systems examined include the NVIDIA Tesla M2070 GPGPU and the Intel Xeon X5650 six-core processor. To ensure a fair comparison the GPGPU and the Xeon processors also evaluated 1,024 synapses per neuron.

As shown under example 1 in Table 10.4, the specialized neural systems consumed much less power and area compared to the current commercial systems. About 12 NVIDIA GPGPUs or 179 Xeon processors were needed to process the network, leading to a power consumption of 2.7 kW and 17 kW respectively. All the specialized neural systems had a power consumption of less than 2 W and a chip area of less than 36 mm^2. The digital memristor system was 2 times more power efficient and 28 % the area of the SRAM based system. The analog memristor system was 16 times more power efficient and 11 % the area of the SRAM based system.

Each specialized neural core processed only 256 specific neurons. Thus each core was active only 100,000 times a second in this example, and the network was processed at the same rate. This leads to the cores being inactive for part of the time if they can process their 256 neurons at a higher rate. A downside of the specialized cores is that they cannot be multiplexed to process other neurons while they are inactive. Thus a larger network processed at a slower rate may have the same overall neurons/s throughput, but would require more specialized cores sitting idle for a longer time between successive iterations.

To test the impact of this, the second example processes a larger network at a slower rate, but it has the same overall neurons/s throughput. Since the NVIDIA GPGPUs and Xeon processors can multiplex among different neurons, the same number of GPGPUs and Xeon processors would be needed as in the first example (see example 2 in Table 10.4).

More specialized neural cores are needed to accommodate the larger number of neurons in example 2, leading to larger chip area and power compared to example 1. As expected, the active times of the specialized cores were very low due to the low iteration rate. With this low activity rate, it is seen that the memristor based systems

Table 10.4 One bits per neuron systems. Both examples are processing 2.56 billions neurons/s

	Example 1: 25,600 neurons					Example 2: 1,706,667 neurons				
	100,000 iterations/s					1,500 iterations/s				
Configuration	Number of chips	Chip area (mm^2)	% active	Power (W)	Power eff. over Xeon	Number of chips	Chip area (mm^2)	% active	Power (W)	Power eff. over Xeon
Memristor Analog (configuration 1)	1	3.7	9.7	0.07	**253,489**	2	248	0.15	0.70	**24,395**
Memristor Digital (configuration 2)	1	9.7	60.8	0.62	**27,546**	2	333	0.91	1.25	**13,633**
SRAM (configuration 3)	1	35.2	60.8	1.13	**15,099**	5	388	0.91	28.02	**607**
NVIDIA M2070	12	529.0	99.2	2,700.00	**6**	12	529	99.2	2,700.00	**6**
Intel Xeon X5650	179	240.0	99.9	17,005.00	**1**	179	240	99.90	17,005.00	**1**

are far more power efficient than the SRAM based neural cores. Since memristors can retain their data when powered off, we assumed that the memristor based cores could be powered down during their inactive periods. However the routers on these chips were assumed to be active throughout, thus leading to higher power consumption for the larger chips in example 2 compared to example 1. The SRAM neural cores had higher energy consumptions due to the leakage energy of the SRAM arrays. The SRAM cores could potentially be set to a low power data retentive mode to increase their power efficiencies (as described in [13, 14]). Further study is needed to examine the impact of this mode on our cores.

Table 10.5 lists the data for the same examples as in Table 10.4, but for systems with 4 bits per neuron and input. In this case the chip areas and power were larger because the cores were bigger and slower than in the 1 bit case. The power efficiency trends however were the same as in example 1.

10.8 Conclusions

This chapter examined the design of several high performance, low power specialized neural cores. These included both memristor and SRAM based cores. Two cases of the memristor cores were studied—digital and analog. Full system performance prediction of multicore processors with these cores and specialized on-chip routers were developed.

The results indicate that specialized multicore processors can provide significant area, power, and speed efficiencies over current commodity high performance computer platforms. These efficiencies come at the cost of generalizability—however a large body of studies has shown that neural processing can be applied a broad range of applications [11, 12, 22]. Memristor based cores provide the highest efficiencies, while in several cases the SRAM based specialized cores were very efficient as well. The SRAM based cores lost efficiency when they were idle for longer periods of time due to leakage energy loss. Specialized low power circuit techniques as utilized by IBM in their SRAM based neural core [13, 14] can drastically cut down on this power. Some of these techniques could be applied to the memristor cores and possibly the on-chip routers examined.

Custom neuromorphic architectures consume extremely low power and can enable deployment of intrusion detection systems across all devices in a network. Future work in this area will be to examine the performance of cyber security applications implemented utilizing the proposed neuromorphic systems.

Table 10.5 Four bits per neuron systems. Both examples are processing 2.56 billions neurons/s

	Example 1: 25,600 neurons					Example 2: 1,706,667 neurons				
	100,000 iterations/s					1,500 iterations/s				
Configuration	Number of chips	Chip area (mm^2)	% active	Power (W)	Power eff. over Xeon	Number of chips	Chip area (mm^2)	% active	Power (W)	Power eff. over Xeon
Memristor Analog (configuration 4)	1	5.9	38.6	0.07	**234,859**	1	395	0.58	0.70	**24,210**
Memristor Digital (configuration 5)	1	18.2	89.6	0.62	**16,968**	4	303	1.34	1.63	**10,419**
SRAM (configuration 6)	1	29.1	89.6	1.13	**8,215**	9	383	1.34	48.67	**349**
NVIDIA M2070	12	529.0	99.2	2,700.00	**6**	12	529	99.2	2,700.00	**6**
Intel Xeon X5650	179	240.0	99.9	17,005.00	**1**	179	240	99.9	17,005.00	**1**

References

1. B. Belhadj, A. J. L. Zheng, R. Héliot, and O. Temam. "Continuous real-world inputs can open up alternative accelerator designs," SIGARCH Comput. Archit. News 41, 3 (June 2013)
2. Steve Esser, Alexander Andreopoulos, Rathinakumar Appuswamy, Pallab Datta, Davis Barch, Arnon Amir, John Arthur, Andrew Cassidy, Myron Flickner, Paul Merolla, Shyamal Chandra, Nicola Basilico, Stefano Carpin, Tom Zimmerman, Frank Zee, Rodrigo Alvarez-Icaza, Jeffrey Kusnitz, Theodore Wong, William Risk, Emmett McQuinn, Tapan Nayak, Raghavendra Singh and Dharmendra Modha, "Cognitive Computing Systems: Algorithms and Applications for Networks of Neurosynaptic Cores," International Joint Conference on Neural Networks, Dallas, Texas, 2013.
3. T. Taha, R. Hasan, C. Yakopcic and M. McLean, "Exploring the Design Space of Specialized Multicore Neural Processors," International Joint Conference on Neural Networks (IJCNN), August 2013. Available at: http://homepages.udayton.edu/~ttaha1/papers/IJCNN13_cores.pdf
4. T. Bass, "Intrusion detection systems and multisensor data fusion,"Communications of the ACM 43, no. 4 (2000): 99-105.
5. E. Marcello, C. Mazzariello, F. Oliviero, S. P. Romano, and C. Sansone, "Evaluating Pattern Recognition Techniques in Intrusion Detection Systems," In PRIS, pp. 144-153. 2005.
6. N. Jiang, and Li Yu, "Intrusion detection using pattern recognition methods," In Optics East 2007, pp. 67730S-67730S. International Society for Optics and Photonics, 2007.
7. L. O. Chua, "Memristor-The Missing Circuit Element," IEEE Transactions on Circuit Theory, vol.18, no.5, pp 507-519, 1971.
8. D. B. Strukov, G. S. Snider, D. R. Stewart, and R. S. Williams, "The missing Memristor found," Nature, 453, pp 80-83, 2008.
9. R. S. Williams, "How We Found The Missing Memristor," IEEE Spectrum, vol. 45, no. 12, pp. 28-35, 2008.
10. C. Yakopcic, T. M. Taha, G. Subramanyam, R. E. Pino, "Generalized Memristive Device SPICE Model and its Application in Circuit Design," IEEE Transactions on Computer-Aided Design of Integrated Circuits and Systems, 23 (2013) pp. 1201-1214.
11. H. Esmaeilzadeh, A. Sampson, L. Ceze, and D. Burger, "Neural Acceleration for General-Purpose Approximate Programs," International Symposium on Microarchitecture (MICRO), 2012.
12. H. Esmaeilzadeh, A. Sampson, L. Ceze, and D. Burger, "Towards Neural Acceleration for General-Purpose Approximate Computing," Workshop on Energy Efficient Design (WEED), 2012.
13. P. Merolla, J. Arthur, F. Akopyan, N. Imam, R. Manohar, D. S. Modha, "A digital neurosynaptic core using embedded crossbar memory with 45pJ per spike in 45nm," IEEE Custom Integrated Circuits Conference (CICC) pp.1-4, 19-21 Sept. 2011.
14. J. V. Arthur, P. A. Merolla, F. Akopyan, R. Alvarez, A. Cassidy, S. Chandra, S. K. Esser, N. Imam, W. Risk, D. B. D. Rubin, R. Manohar, D. S. Modha, "Building block of a programmable neuromorphic substrate: A digital neurosynaptic core," International Joint Conference on Neural Networks (IJCNN), pp.1-8, June 2012.
15. N. Muralimanohar, R. Balasubramonian, and N. Jouppi, "Optimizing NUCA Organizations and Wiring Alternatives for Large Caches with CACTI 6.0" In Proceedings of the 40th Annual IEEE/ACM International Symposium on Microarchitecture (MICRO 40), Washington, DC, USA, 3-14, 2007.
16. A. B. Kahng, B. Li, L. S. Peh, and K. Samadi, "ORION 2.0: A fast and accurate NoC power and area model for early-stage design space exploration," Design, Automation & Test in Europe Conference & Exhibition, pp.423-428, 20-24 April 2009.
17. W. Lu, K.-H. Kim, T. Chang, S. Gaba, "Two-terminal resistive switches (memristors) for memory and logic applications," Asia and South Pacific Design Automation Conference, pp. 217-223, 2011.

18. C. Yakopcic and T. M. Taha, "Energy Efficient Perceptron Pattern Recognition Using Segmented Memristor Cross-bar Arrays," IEEE International Joint Conference on Neural Networks (IJCNN), August 2013.
19. S. Shin, K. Kim, S.-M. Kang, "Analysis of Passive Memristive Devices Array: Data-Dependent Statistical Model and Self-Adaptable Sense Resistance for RRAMs," Proceedings of the IEEE, 100(6), June 2012.
20. B. Han, and T. M. Taha, "Acceleration of spiking neural network based pattern recognition on NVIDIA graphics processors," Journal of Applied Optics, 49(101), pp. 83-91, 2010.
21. J. M. Nageswaran, N. Dutt, J. L. Krichmar, A. Nicolau, and A.Veidenbaum, "Efficient simulation of large-scale spiking neural networks using CUDA graphics processors," In Proceedings of the 2009 international joint conference on Neural Networks (IJCNN). IEEE Press, Piscataway, NJ, USA, 3201-3208, 2009.
22. T. Chen, Y. Chen, M. Duranton, Q. Guo, A. Hashmi, M. Lipasti, A. Nere, S. Qiu, M. Sebag, O. Temam, "BenchNN: On the Broad Potential Application Scope of Hardware Neural Network Accelerators," IEEE International Symposium on Workload Characterization (IISWC), November 2012.

Chapter 11
Memristor SPICE Model Simulation and Device Hardware Correlation

Robinson E. Pino, Antonio S. Oblea, and Kristy A. Campbell

11.1 Introduction

The memristor device was first described in 1971 by Leon Chua [1] as the fourth basic circuit element. Recently, the memristor has received much attention since the publication of the paper titled "The missing memristor found" in 2008 describing the memristive characteristics of metal-oxide-based memristor devices [2]. The memristor name is a contraction for memory resistor [1]. It is a two terminal passive device whose resistance state depends on its previous state. Given their two terminal structural simplicity and electronic passivity, the applications for memristor technology range from non-volatile memory, instant on computers, reconfigurable electronics and neuromorphic computing [3, 4]. Several device models have been presented to describe the electrical behavior of memristor devices [1, 2, 4–6]. However, there is no paper to the best of our knowledge in the published literature that shows model versus hardware plot correlations within a SPICE microelectronics industry standard environment. Recently, we developed an empirical model that accurately describes the electrical behavior of ion-conductor chalcogenide-based memristors [7]. In this work, we present a SPICE-based version of our memristor device compact model.

R.E. Pino (✉)
U.S. Department of Energy, Office of Science, Washington, DC 20585, USA
e-mail: robinson.pino@science.doe.gov

A.S. Oblea • K.A. Campbell
Department of Electrical and Computer Engineering, Boise State University, Boise, ID 83725, USA
e-mail: kriscampbell@boisestate.edu

R.E. Pino et al. (eds.), *Cybersecurity Systems for Human Cognition Augmentation*, Advances in Information Security 61,
DOI 10.1007/978-3-319-10374-7_11

11.2 Experimental Details

The fabrication and electrical characterization of our characterized memristor devices have been described elsewhere [8]. In summary, the devices were fabricated on 200 mm p-type Si wafers. The device structure consists of 180 nm diameter layers of 300 Å Ge2Se3/500 Å Ag2Se/100 Å Ge2Se3/500 Å Ag/100 Å Ge2Se3 contacted by 80 μm^2 tungsten contacts for both top and bottom node contacts. Electrical measurements were performed with an Agilent B1500A semiconductor parameter analyzer. Figure 11.1 displays the characteristic memristive Lissajous I–V electrical responses under sinusoidal input voltage of 0.5 V amplitude at three operating frequencies of 10, 100, and 1,000 Hz respectively. From the figure, we can observe that the memristive non-linear I–V characteristics narrows as the frequency increases from 10 to 1,000 Hz as originally postulated by Chua in 1971.

11.3 SPICE Compact Model

The basic mathematical definition of a memristor is that of a voltage-controlled, one-port device in the generalized class of nonlinear dynamical systems called memristive systems described by (11.1) and (11.2) where x can be a set of state variables and f is a generalized function [4]. Also, G is the conductance whose reciprocal is the resistance, R, and the variables time, current, and voltage are represented by t, i, and v.

$$i = G(x, v, t)v \tag{11.1}$$

$$dx/dt = f(x, v, t) \tag{11.2}$$

Our SPICE model treats the memristor device as a voltage-controlled current source simplifying (11.1) to a function $i = G(v)v = v/R(v)$. Also, it describes the resistance as function of the voltage across the device, $R(v)$, schematically shown in Fig. 11.2. The complete SPICE model netlist, including all fitting parameter values, is shown in Table 11.1.

Figure 11.2a shows the memristor element symbol, and (b), the schematic compact model representation modeled as a series combination of two voltage-controlled resistors in series, Rhigh and Rlow. In particular, Rhigh describes the highest resistance and nonlinear dynamic switching to the lowest resistance state Ron. The latter, Rlow, describes the lowest resistance and nonlinear dynamic return to the highest resistance state ROff. In Table 11.1, the memristor device effective resistance value is described by the non-linear piece-wise functions Rh(v) and Rl(v). Rh(v) describes the memristance behavior by an exponential decay for an increasing input voltage greater than the threshold voltage, Th. For voltages less than Th, the output resistance equals Roff. Otherwise, the model output for Rhigh is set 1 Ω. Rh1, Rh2, Rh3, and Vh are empirical parameters used to capture the nonlinear dynamic transition

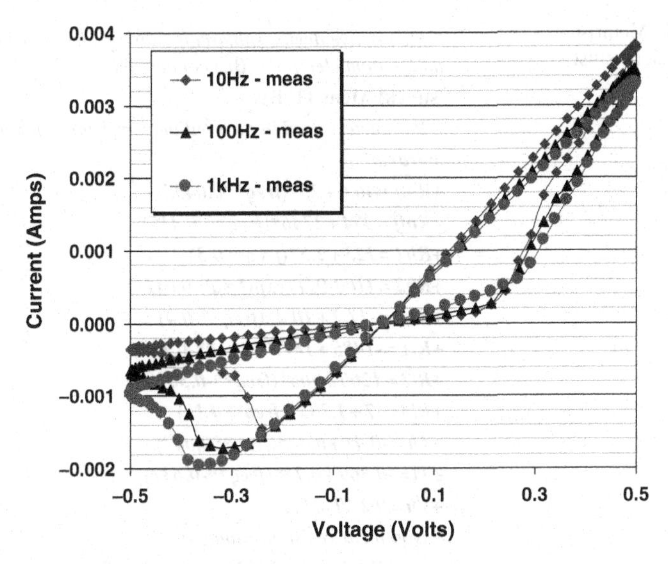

Fig. 11.1 Typical memristor device Lissajous I–V characteristic behavior for 0.5 V amplitude sinusoidal inputs at 10, 100, and 1,000 Hz

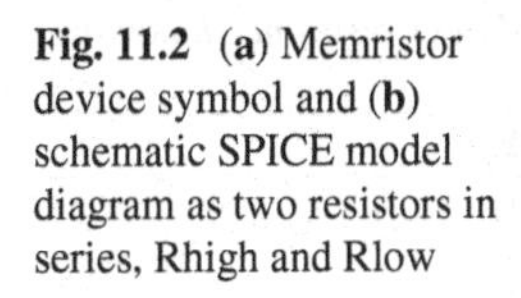

Fig. 11.2 (**a**) Memristor device symbol and (**b**) schematic SPICE model diagram as two resistors in series, Rhigh and Rlow

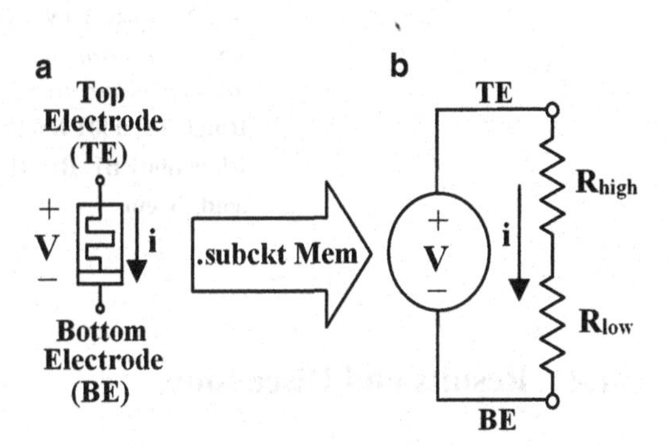

characteristics of the memristor device. On the other hand, Rl(v) describes the memristance behavior by an exponential growth for a decreasing input voltage less than the threshold, Tl, and for decreasing input voltages greater than Tl, the resistance is set to Ron. Otherwise, Rl(v) is set to 1 Ω resistance. Similarly, Rl1, Rl2, Rl3, and Vl are empirical parameters used to capture the nonlinear transition characteristics of the memristor device.

During operation, the SPICE compact model implementation showed in Table 11.1 works by having only one of the resistors limit the current flow through the device at any given time. For example, in the low-resistance state, Rlow is set to Ron = 46.6 + 75 × (frequency of operation)0.0435 Ω and Rhigh is set to 1 Ω (SPICE could potentially have convergence issues when a resistor value is zero). Conversely, in the high-resistance state, Rhigh is set to Roff = 394 + 2,746 × (frequency of operation) − 0.415 Ω and Rlow is set to 1 Ω as shown in detail in Table 11.1 respectively.

Table 11.1 Memristor SPICE Model Netlist

* Memristive device subcircuit with top electrode (TE) and bottom electrode (BE) connections
.subckt Mem TE BE
* Parameters used in the nonlinear resistance functions
.params
+Ron =46.6+75*(freq**0.0435)
+Roff=394+2746*(freq**-0.415)
+Rh1=123+2.5*(freq**0.3)
+Rh2=1100-911*exp(freq*-0.011)
+Rh3=-17.2+10.4*(freq**-0.4)
+Rl1=-1485+1677*(freq**-0.115)
+Rl2=1200+400*(freq**-0.30103)
+Rl3=-7+1.333*(freq**0.18)
+Th=-0.41+0.48*(freq**0.047)
+Tl=-0.369+0.73*(freq**-0.637)
+Vh=0.2 Vl=-0.5
* Nonlinear resistance functions
.func Rh(V)=IF (ddt(V)>=0 & V>Th, Rh1 +
+ Rh2*exp(Rh3*(V - Vh)), IF(ddt(V)>=0, Roff, 1))
.func Rl(V)=IF (ddt(V)<0 & V<Tl, Rl1 +
+ Rl2*exp(Rl3*(V - Vl)), IF(ddt(V)<0, Ron, 1))
* Series resistors that use the nonlinear resistance functions to model memristive behavior
Rhigh TE n001 R=Rh(V(TE,BE))
Rlow n001 BE R=Rl(V(TE,BE))
.ends Mem

11.4 Results and Discussion

To verify the operation of the SPICE memristor model, simulations were performed in LTSpice (free-ware SPICE simulator) and compared to the physical device data from Fig. 11.1 [8]. Figure 11.3 shows the correlation between the hardware electrical characterization data and the memristor model output. The compact model fitting parameters to achieve the model fit are tabulated in Table 11.1.

In addition, Fig. 11.4 displays the time-dependent SPICE model results versus hardware electrical characterization data. The data is for an operating 0.5 V amplitude sinusoidal input signal at a frequency of 1,000 Hz. It is important to point out that our SPICE model is parameterized as function of frequency. Thus, the netlist in Table 11.1 describes describe the memristor devices operation as function of frequency.

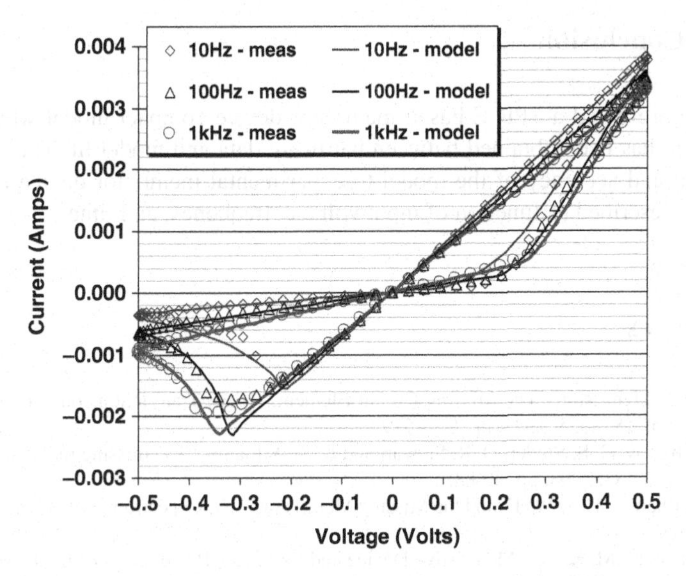

Fig. 11.3 Memristor hardware, markers, versus SPICE model correlation at three operating input frequencies of 10, 100, and 1,000 Hz

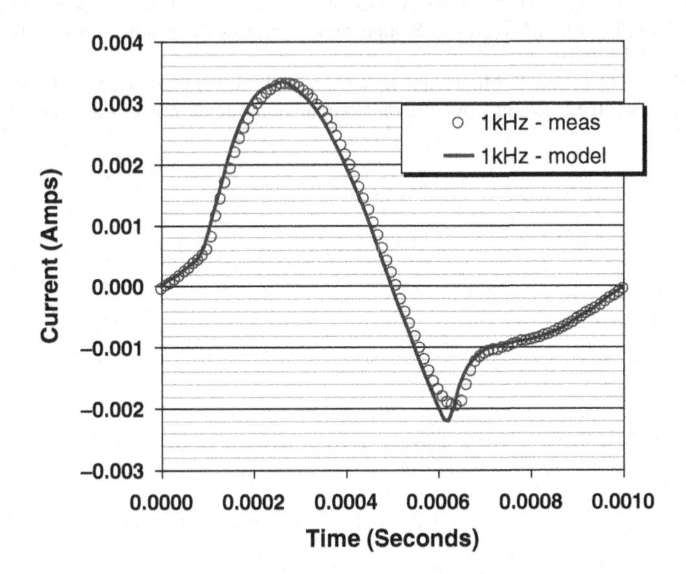

Fig. 11.4 Memristor SPICE model output current, *line*, versus device hardware data, *circles*, fit as function of time for an input 0.5 V sinusoidal bias at 1,000 Hz

11.5 Conclusion

We have presented a SPICE-based memristor device compact model where good agreement has been obtained between hardware data and model fit. The electrical operation and accuracy of the model to experimental memristor device hardware has been described as function of input voltage, frequency, and time.

References

1. L. Chua, "Memristor - The Missing Circuit Element," IEEE Transactions on Circuits Theory (IEEE), vol. 18, no. 5, 1971, pp. 507–519.
2. D. B. Strukov, G. S. Snider, D. R. Stewart and R. S. Williams, "The missing memristor found," Nature, vol. 453, 2008, pp. 80-83.
3. R. S. Williams, "How We Found the Missing Memristor," IEEE Spectrum, vol. 45, no. 12, 2008, pp. 28-35.
4. L. Chua and S.M. Kang, "Memristive Device and Systems," Proceedings of IEEE, Vol. 64, no. 2, 1976, pp. 209-223.
5. Z. Biolek, D. Biolek, V. Biolková, "Spice Model of Memristor With Nonlinear Dopant Drift," Radioengineering, vol. 18, no. 2, 2009, pp. 210-214.
6. Y. N. Joglekar and S. J. Wolf, "The elusive memristor: properties of basic electrical circuits," European Journal of Physics, vol. 30, 2009, pp. 661–675
7. R.E. Pino, J.W. Bohl, N. McDonald, B. Wysocki, P. Rozwood, K.A. Campbell, A.S. Oblea, and A. Timilsina, "Compact Method for Modeling and Simulation of Chalcogenide Based Memristor Devices," IEEE/ACM International Symposium on Nanoscale Architectures 2010, Anaheim, CA, June 17-18, 2010.
8. A.S. Oblea, A. Timilsina, D. Moore, and K.A. Campbell, "Silver Chalcogenide Based Memristor Devices," 2010 IEEE World Congress on Computational Intelligence, Barcelona, Spain, July 18-23, 2010.

Chapter 12
Reconfigurable Memristor Based Computing Logic

Robinson E. Pino and Youngok K. Pino

12.1 Introduction

Reconfigurable computing logic describes the ability to transform the functionally of a Boolean function from let us say a logic AND gate to a logic OR gate functionality without physical rewiring and vice versa. Today, this type of logic reconfiguration is not possible as the operational functionality of transistors, resistors, capacitors, and inductors is fixed and electronically unchangeable. However, there exists a new electronic device whose impedance state is electronically variable and non-volatile. The name of such device is the memristor. The electronic operational and behavioral characteristics of memristor devices have been reported recently in the literature by the authors [1, 2]. Pino and Bohl have described mathematically that memristors can operate within a range of impedance states bounded by a maximum, RHigh, and minimum, RLow, resistance values. The switching characteristics of the memristor devices between their ON and OFF states, RHigh and RLow, are governed by discrete threshold voltages, VHigh and VLow, that switch the device ON and OFF respectively [1]. In addition, the memristor device whose name stands for memory resistor is of particular interest because it is a passive device that when power is turned off, it remembers its previous impedance state [3–5]. In this work, we make use of the memristor device as a memory element within our reconfigurable computing logic architecture.

R.E. Pino (✉)
U.S. Department of Energy, Office of Science, Washington, DC 20585, USA
e-mail: robinson.pino@science.doe.gov

Y.K. Pino
Information Sciences Institute, University of Southern California, Arlington, VA 22203, USA
e-mail: ypino@isi.edu

R.E. Pino et al. (eds.), *Cybersecurity Systems for Human Cognition Augmentation*, Advances in Information Security 61,
DOI 10.1007/978-3-319-10374-7_12

12.2 Computing Architecture Design

Our innovative reconfigurable digital computing architecture design consists of several key elements. A digital binary decoder is used to take multiple inputs and select a single output out of multiple combinations. A binary decoder which has an n-bit binary input code and a one activated output out-of-2n output code is called a binary decoder, and it is used when it is necessary to activate exactly one of the 2n outputs based on an n-bit input value [6]. The remaining computing architecture design components are: n- and p-channel field effect transistors, resistors, and inverting logic gates. Figure 12.1 describes the design architecture of a single binary input reconfigurable inverting/non-inverting logic gate. The function of node VC is to deliver a power to the circuit, node VR delivers the threshold voltages required to switch the memristors M1 and M2 ON/OFF, and node K is the select 1/0 to operate/configure the reconfigurable logic gate respectively. From the figure, we can see that when for example the decoder input is 0, the decoder will select the output line A1. This means that the output of the decoder will be A1 = 1 and A2 = 0. In this instance, the gate of n-channel transistor T5 will be selected turning the transistor to the ON state. Similarly, n-channel transistor T8 will remain in the OFF state as the output of A2 is currently 0. Assuming the signal at nod K = 0, for operation of the logic gate, the p-channel transistors T1 and T3 will be on their ON state while the n-channel transistors will be in the OFF state. Thus, when K = 0, the VC will be connected to the drain terminal of transistors T5 and T8. Also, the memristors will be connected to ground through the p-channel transistor T3. This biasing scheme will allow the VC node to be connected in series with resistor R1 and in parallel with transistors T5 and T8 in turn whose source terminals are connected to memristors M1 and M2 respectively.

Assuming that we can neglect the channel resistance for all transistors used in Fig. 12.1, the voltage across the memristor M1 and M2 are given by the following two equations

$$V\left(M1\right) = M1\frac{V_C}{R1 + M1} \quad \mathit{if}\, A1 = 1 \;\mathit{else}\; V\left(M1\right) = 0 \tag{12.1}$$

$$V\left(M2\right) = M2\frac{V_C}{R2 + M2} \quad \mathit{if}\, A2 = 1 \;\mathit{else}\; V\left(M2\right) = 0 \tag{12.2}$$

From the electronic properties of memristor devices [2], we can assume that the impedance variation can vary from RLow = 100 Ω up to RHigh = 2,000 Ω. For an R1 resistance value of 250 Ω, the voltage across the memristor can vary from V(M = RLow) = 0.29 V up to V(M = RHigh = 0.89 V. Then since A1 = 1 and K = 0, both the n- and p-channel transistors T6 and T7 will be on the ON state respectively. This will allow the voltage to at the input of the inverting circuit Inv1 to range between 0.29 and 0.89 V. The function of the inverter circuit as the name suggests is to invert the value input to it. For example, if the input to the inverter is 0 the output will be 1 and vice versa. In addition, an ideal inverter circuit can be described

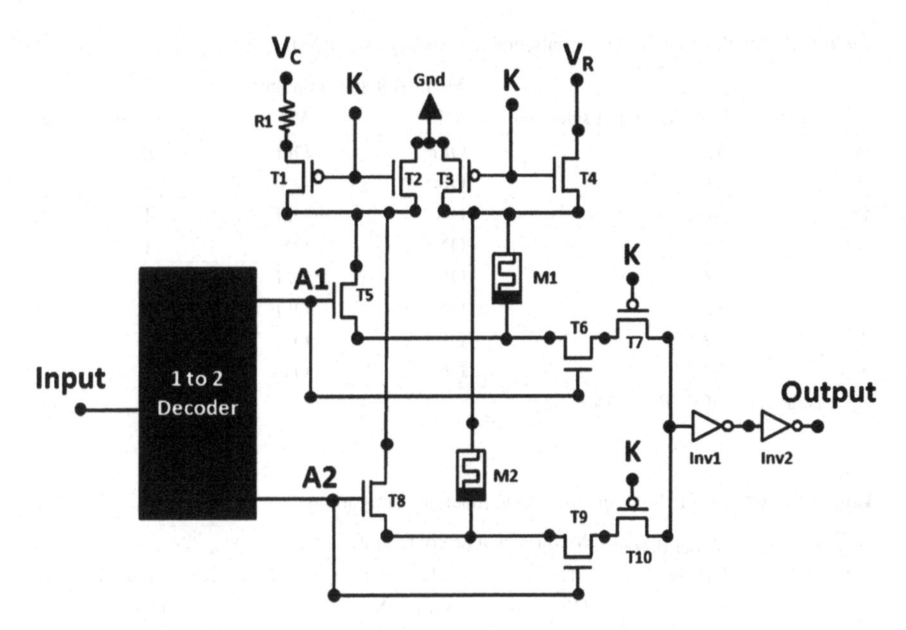

Fig. 12.1 Single binary input reconfigurable logic design

by a sigmoid threshold function in which the threshold for determining the output of the circuit corresponds to half the operating voltage. Thus, for the inverter circuit to output 1 V, the input voltage needs to be below 0.5 V if the operating voltage is 1 V, and if the inverter output of the inverter is to be 0 V, the input to the inverter circuit needs to be above 0.5 V. Then, if we look at the voltage range across the memristor it is from 0.29 to 0.89 V. Therefore, when the voltage across the memristor is 0.29 V, the output of inverter Inv1 will be 1 V, and when the input is 0.89 V, the output of the inverter will be 0 V. Given that the output of the first inverter Inv1 will output the complement of computing logic function, a second inverter Inv2 is used to correct the logic output. In summary, Table 12.1 shows all the possible combinations in which the various ON/OFF memristor values combinations will reconfigure the logic output. If we look closely at the circuit design in Fig. 12.1 and the results in Table 12.1, we can see that there is one memristor for each decoder output, and the output is directly proportional to the state of the memristor connected to it. Thus the overall output of the logic circuit depends on a single memristor device that is chosen through the decoder, and the state of any other memristor can be ignored as shown in Table 12.1. In Fig. 12.1, the p-channel transistors T7 and T10 are used to decouple the unselected memristor as the voltage of each unselected decoder output is 0 V. Table 12.2 shows the circuit logic configuration for the design in Fig. 12.1 to exhibit the logic properties of the inverter circuit.

Finally, configuration of the various memristors to achieve logic reconfiguration can be performed when in the circuit schematic design shown in Fig. 12.1, the node K is set to 1 V. This will cause the n-channel transistor T4 to turn ON and the

Table 12.1 Single binary input reconfigurable inverting/non-inverting logic gate

Input value	Decoder output selection	Memristor state configuration		Output value
		M1	M2	
0	A1	OFF	OFF	0
0	A1	OFF	ON	0
0	A1	ON	OFF	1
0	A1	ON	ON	1
1	A2	OFF	OFF	0
1	A2	ON	OFF	0
1	A2	OFF	ON	1
1	A2	ON	ON	1

Memristor state: OFF = R_{Low}, ON = R_{High}

Table 12.2 Computing logic Inverter, INV, function configuration

Input value	Decoder output selection	Memristor state configuration		Output value	Logic function
		M1	M2		
0	A1	ON	OFF/ON	1	INV
1	A2	OFF/ON	OFF	0	

p-channel transistors T1, T3, T7, and T10 to turn OFF. Thus, a direct path from node V_R to ground is now established and node V_C is isolated. Now that the memristor is connected directly to ground, node VR can be use to apply the specific voltage value to switch the memristor device from ON to OFF state and vice versa. Also, only one memristor can be reconfigure at a time, and the memristor that can be reconfigured corresponds to the selected by the output of the decoder. Thus, any particular logic function can be configured one memristor at a time.

As we seek to scale this reconfigurable logic computing architecture, Fig. 12.2 shows how we can use defined cell blocks to simplify the circuit schematic. From the figure, the cell block diagram PwRr is used to power and reconfigure the logic, MBMn represents the nth memristor element, and AMPn the nth output of the reconfigurable logic circuit. Figure 12.3 shows the simplified block cell single binary input reconfigurable circuit logic design.

A two-input one-output reconfigurable computing logic function can be designed as shown in Fig. 12.4. In addition, Tables 12.3 and 12.4 show how a memristor configuration can be set to enable the two-input one-output logic circuit to exhibit the characteristic properties of an AND and XOR computing logic functionalities.

Similarly, Figs. 12.5 and 12.6 show how multi-input multi-output reconfigurable logic functions can be designed by increasing the decodes input/output capacity, adding MBM memristor cell blocks, PwRr, and AMP cell block circuitry. In particular, Fig. 12.6 shows how a multi-output logic function can be designed by adding and additional PwRr and AMP cell block circuitry for each additional desired output.

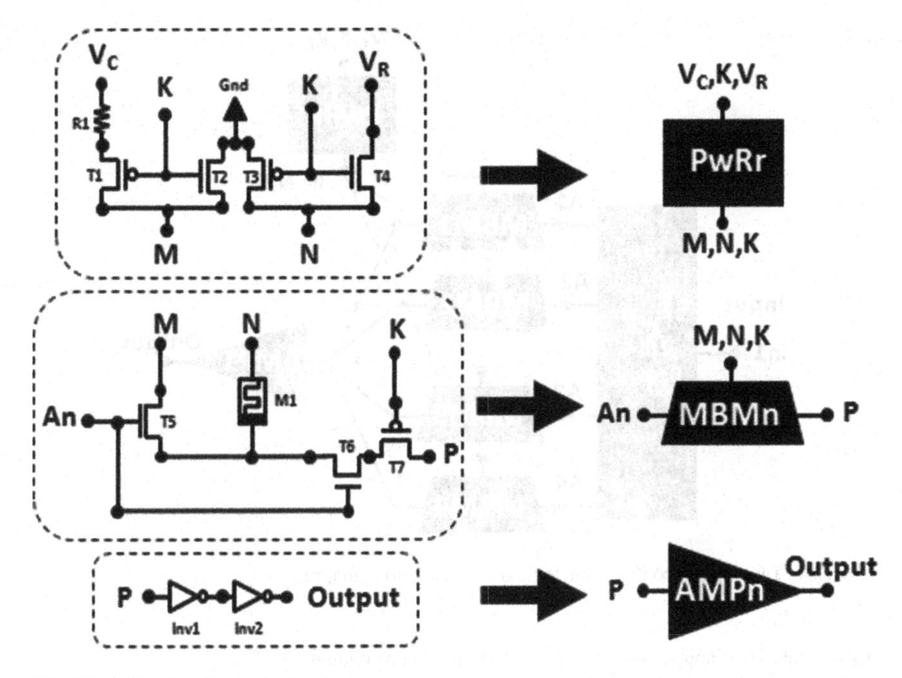

Fig. 12.2 Block cell schematic diagrams

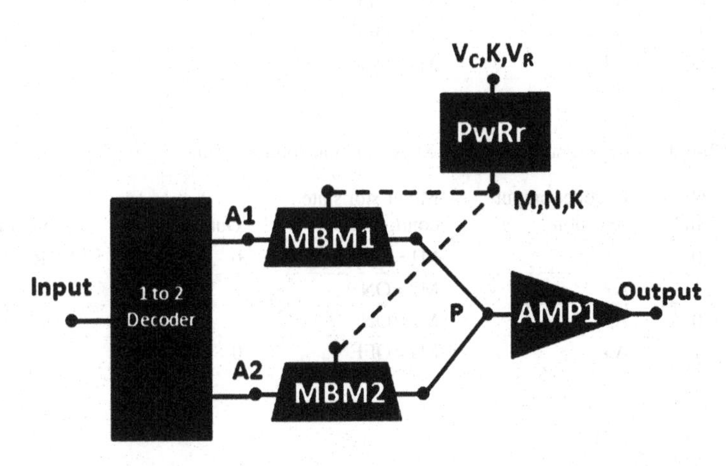

Fig. 12.3 Simplified block cell single binary input reconfigurable logic design

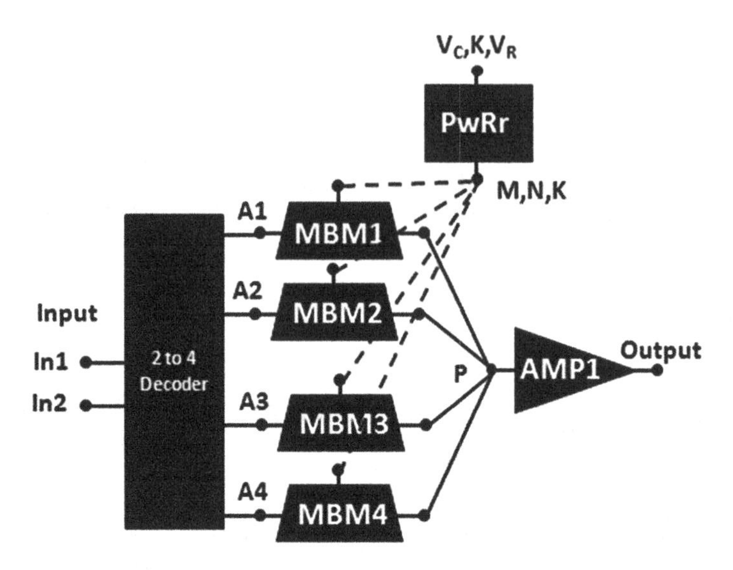

Fig. 12.4 Two input reconfigurable logic design cell block diagram

Table 12.3 Two-input computing logic AND function configuration

Input value		Decoder output selection	Memristor state configuration	Output value	Logic function
0	0	A1	M1 = OFF	0	AND
0	1	A2	M2 = OFF	0	
1	0	A3	M3 = OFF	0	
1	1	A4	M4 = ON	1	

Table 12.4 Two-input computing logic XOR function configuration

Input value		Decoder output selection	Memristor state configuration	Output value	Logic function
In1	In2				
0	0	A1	M1 = OFF	0	XOR
0	1	A2	M2 = ON	1	
1	0	A3	M3 = ON	1	
1	1	A4	M4 = OFF	0	

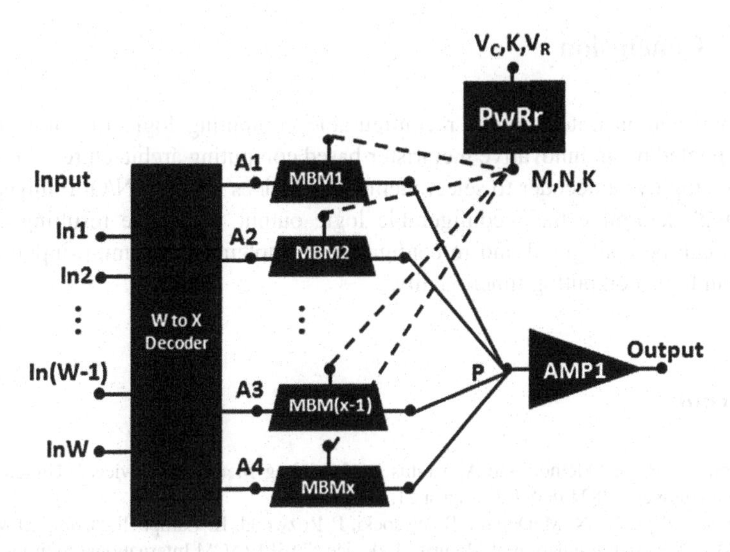

Fig. 12.5 Scalable multi-input reconfigurable logic design cell block diagram

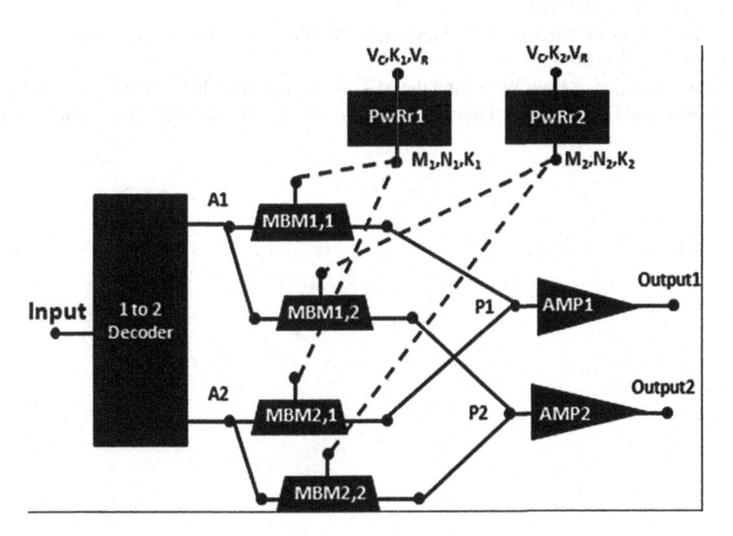

Fig. 12.6 Scalable multi-output reconfigurable logic design cell block diagram

12.3 Conclusion

We have demonstrated how a reconfigurable computing logic function can be implemented by an innovative memristor based computing architecture. The architecture employs a decoder to select memristor devices whose ON/OFF impedance state will determine the reconfigurable logic output. Thus, the resulting circuit design can be configured and re-configured to implement any multi-input/output Boolean logic computing functionality.

References

1. R. Pino, J. Bohl, "Method and Apparatus for Modeling Memristor Devices," United States Patent Number US8249838 B2, August 21, 2012.
2. R. Pino, J.W. Bohl, N. McDonald, B. Wysocki, P. Rozwood, K. Campbell, "Compact Method for Modeling and Simulation of Memristor Devices," IEEE/ACM International Symposium on Nanoscale Architectures, NANOARCH'10, Anaheim, CA, June 17-18, 2010
3. Leon O. Chua, "Memristor - The Missing Circuit Element" IEEE Transactions on Circuits Theory, 18 (1971) 507-519.
4. D. B. Strukov, G. S. Snider, D. R. Stewart, S. R. Williams, "The missing memristor found," Nature, 453 (2008) 80-83.
5. R. Stanley Williams, "How We Found the Missing Memristor," IEEE Spectrum (2008) 29-35.
6. A. P. Godse and D. A. Godse, "Digital Logic Design," Technical Publications Pune, Pune, India (2008).

Chapter 13
Cyber Security Considerations for Reconfigurable Systems

James Lyke and Arthur Edwards

The class of reconfigurable systems, which include the digital field programmable gate array (FPGA) and emerging new technologies such as neuromorphic computation and memristive devices, represent a type of frontier for cyber security. In this chapter, we provide a brief sketch of the field of reconfigurable systems, introduce a few basic ideas about cyber security, and consider the implications of cyber security as it applies to present and future devices. We also attempt to provide some insights on how to add robustness to reconfigurable systems technologies.

13.1 Introduction to Reconfigurable Systems

One definition of reconfigurable systems is "software-defined hardware". In a field programmable gate array (FPGA), for example, it is possible to define almost arbitrary configurations of logic, memory, and interconnections through the use of a bitstream (a binary sequence of ones and zeros) that personalizes an otherwise uncommitted silicon fabric. More generally, a reconfigurable system is a system containing a number of software-defineable "knobs" that can be altered through a configuration mechanism, such as a finite state machine connected to an access port interface. It is a special case of a programmable system, which also includes the class of one-time programmable (OTP) devices (for which incremental programming is possible, but reconfiguration is not). Examples of one time programmable devices include programmable read-only memories (PROMs), anti-fuse FPGAs, and some types of write-able digital video discs.

J. Lyke (✉) • A. Edwards
U.S. Air Force Research Laboratory, Kirtland, NM 87117, USA
e-mail: james.lyke.2@us.af.mil; arthur.edwards@us.af.mil

R.E. Pino et al. (eds.), *Cybersecurity Systems for Human Cognition Augmentation*, Advances in Information Security 61,
DOI 10.1007/978-3-319-10374-7_13

13.1.1 Motivation/Benefits of Reconfigurability

Most of the benefits of reconfigurable systems can be reduced to one or more of three basic possibilities. In the first case, reconfigurability provides the possibility of "tele-alteration", which is simply the ability to alter in situ the intrinsic function of the circuit. There are several reasons why tele-alteration is desirable:

- Correction of defects. Even the best software code is laden with defects, at the rate of about one part per thousand. Since complex digital designs are often expressed in a high order design language (such as VHDL), it is reasonable to conclude that complex designs may also be laden with defects. Such defects are costly to rectify after fabrication in custom application-specific integrated circuits (ASICs), but may be impossible altogether in fielded systems, especially spacecraft and other inaccessible platforms.
- Time-sharing limited resources. At one level, a reconfigurable system can be thought of as a set of resources that can be dynamically repurposed by reconfiguration. If for example a 5,000 gate FPGA needs to implement 10,000 gates of functionality, separable into two 5,000 gate blocks that are also separated in time, then it is possible in principle to reconfigure the FPGA dynamically. In this case, the FPGA does 10,000 gates "worth of work" with only 5,000 gates. While there are many limitations with this contrived example, it serves to illustrate the potential benefit.
- Adapting to new requirements. Reconfigurable systems are uniquely capable of undergoing wholesale replacement of core functions in systems that may already be fielded.

The second benefit of reconfigurability is a potential enhancement in robustness and resilience through alteration. One example of this is the self-healing wiring harness [1], based on the notion of circumlocution of defective wires (i.e., re-patching in a different set of unused wires to replace a faulty set).

The third benefit of reconfigurabilty is "x-on-demand". Reconfigurability allows for blank copies to be retrieved from storage and programmed rapidly. In reconfigurable concepts such as FPGAs, pre-built parts are personalized rapidly, averting the long fabrication cycle of a dedicated chip.

13.1.2 Taxonomy Reconfigurable Systems

While the FPGA is the best known example of a reconfigurable system, it is only a subset of a much broader class of possibilities as reflected in the taxonomy shown in Fig. 13.1. Traditional stored program (i.e., von Neumann) computers are included in this categorization, since the execution of software programs involves manipulating the contents of an instruction register in a particular sequence to execute a particular algorithm. Of course, contemporary computers are far more sophisticated, having many configurable memory structures and programmable features.

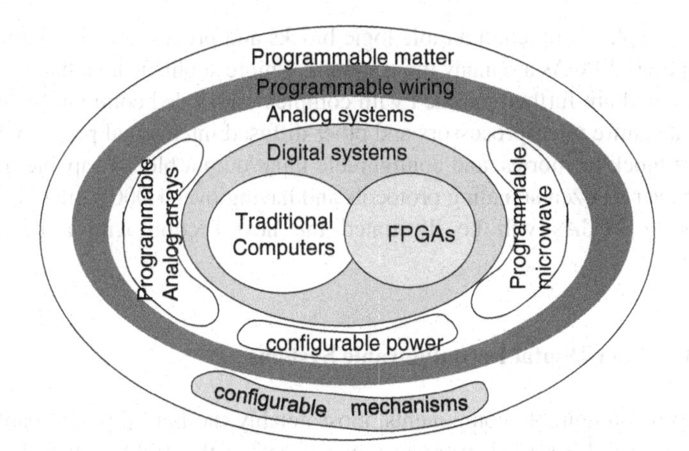

Fig. 13.1 Taxonomy of reconfigurable systems

Figure 13.1 itself, however, extends well beyond digital systems. It includes the possibility that analog, microwave, and mechanical systems could be software defined. It recognizes that the pathways and conduits bridging components together can also be reconfigured. In this section, we briefly survey some elements consistent with these broader ideas of reconfigurable systems.

13.1.2.1 Traditional Computers

In the theory of computation, all computers are equivalent and can be mapped to a model known as a Turing machine. Even though the resource-constrained microcontroller running a wristwatch seems quite different from a multicore supercomputer, they are, from the standpoint of computational complexity theory, equivalent because they can be mapped into a Turing machine formalism. The same complexity theory holds that all algorithms can be executed on Turing machine and therefore can be executed on any computer, no matter how simple [2]. The execution of algorithms involves delivering a sequence of instructions in a defined order, as represented in some binary storage structure that holds their encoded representation. Hence, even the simplest computer is a reconfigurable system.

13.1.2.2 Field Programmable Gate Arrays (FPGAs)

FPGAs evolved from rudimentary programmable logic arrays (PLAs), originally used to implement small numbers of customized Boolean functions by directly encoding logic truth tables. This was achieved by selectively breaking links in a two level logic structure (usually by burning out resistive fuses) to form small amounts of Boolean combinational logic. PLA's became more sophisticated and were eventually replaced by configurable logic blocks and programmable routing grids. Contemporary FPGAs are many thousand times more sophisticated than these early examples, and are further enriched with complex embedded configurable modules (to include entire microprocessors and other diffused intellectual property blocks), multiport block memories, and configurable input/output blocks capable of implementing over a dozen signaling protocols and having over 1,000 signal pins. Some concepts of FPGAs will be illustrated the next section with a pedagogical example.

13.1.2.3 Other Digital Reconfigurable Systems

Specialty reconfigurable components, most notably the field programmable interconnect device (FPID) [3], were first introduced in the 1990s. An FPID can be thought of simply as a routing-only version of a field programmable gate array. Due to the limitations of FPIDs (they generally were only capable of implementing digital signals) and the growing capability of FPGAs (often more than capable of implementing the same functions), their popularity diminished, and the companies that introduce them (such as Aptix and iCube) disappeared from the marketplace.

13.1.2.4 Analog Systems

While even today, most FPGAs are "digital only" systems, digital circuitry itself can be thought of as a special case of analog systems. We can consider analog, therefore, as a broader class of reconfigurable systems, containing configurable circuit elements that carry analog (to include power and microwave) signals and implement discretized (sampled) or continuously variable analog functions.

One of the simplest analog building blocks is the resistor. As shown in Fig. 13.2a, when implemented as a transistor switch, a resistor can have two values, one being a finite imprecise value and the other being effectively infinity. Through a simple circuit, such as a binary resistance ladder, with transistors to short circuit selected resistors (in this case, the transistors are assumed to be perfect zero resistance switches), 16 different resistance values can be selected approximating a reconfigurable resistor (Fig. 13.2c). Using memristive materials, it is possible to replace the more sophisticated binary resistive chain in Fig. 13.2b with single piece of configurable resistive material.

The ability to infuse building blocks such as the programmable resistor within an integrated circuit leads to a lot of interesting possibilities. For example, it is possible to form summing junctions, in which a number of variable resistors can be combined. One application of the summing junction is the approximation of an artificial neuron, as suggested in Fig. 13.3. In the perceptron model of the artificial neuron, the output of a neuron is a thresholded representation of the weighted sum of its inputs. The situation corresponds very approximately to a summing junction network formed by variable resistors (as shown), which serve the role of weights. An

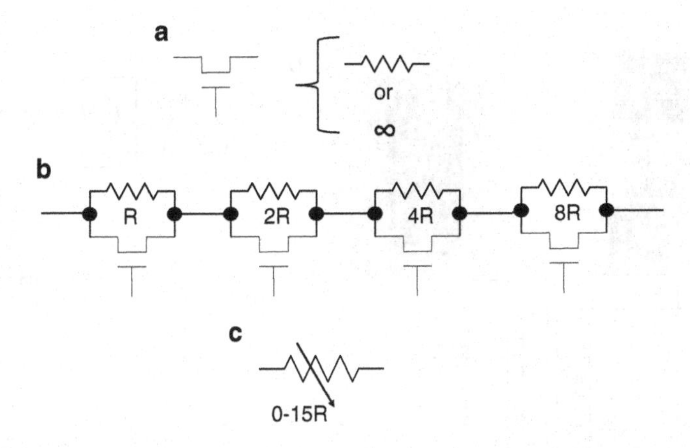

Fig. 13.2 Approximations of programmable resistance. (**a**) Binary setting. (**b**) A configurable binary chain. (**c**) A variable resistance

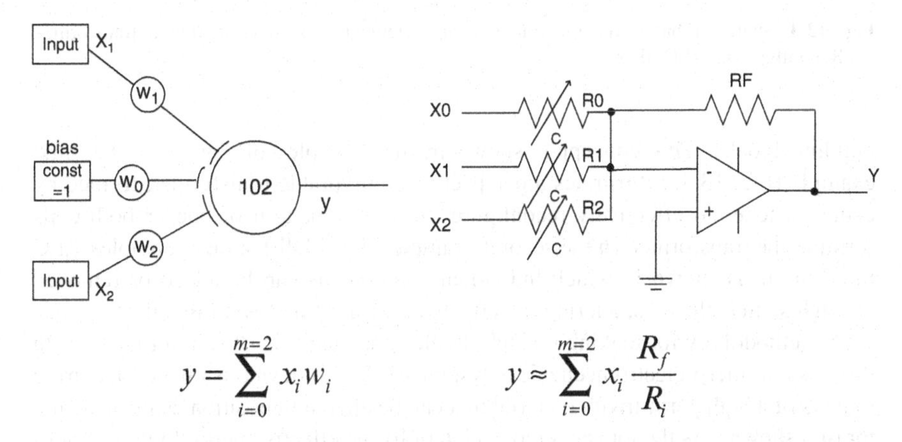

$$y = \sum_{i=0}^{m=2} x_i w_i \qquad y \approx \sum_{i=0}^{m=2} x_i \frac{R_f}{R_i}$$

Fig. 13.3 Approximation of a configurable artificial neuron (perceptron) (*left*) with analog circuit containing several programmable resistors

idealized operational amplifier is used to threshold the voltage formed by the summing network. In this case, which corresponds to the so-called linear combiner [4], when the sum of input exceeds zero volts, the output of the neuron is the maximum voltage corresponding to saturation of the operational amplifier (corresponding to Boolean logic state "1"). Otherwise, the output voltage of the neuron corresponds to the maximum negative (for differential power supplies) or zero voltage (in the case of single ended power supplies), which corresponds to Boolean logic state "0".

In addition to programmable resistances, it is straightforward to imagine programmable capacitance, configurable inductance, and field programmable transistor arrays. These configurable circuit elements can be realized in a number of ways. As in the Fig. 13.2b case for programmable resistance, the strategic infusion of configurable switching elements can be used to approximate reconfigurable analog

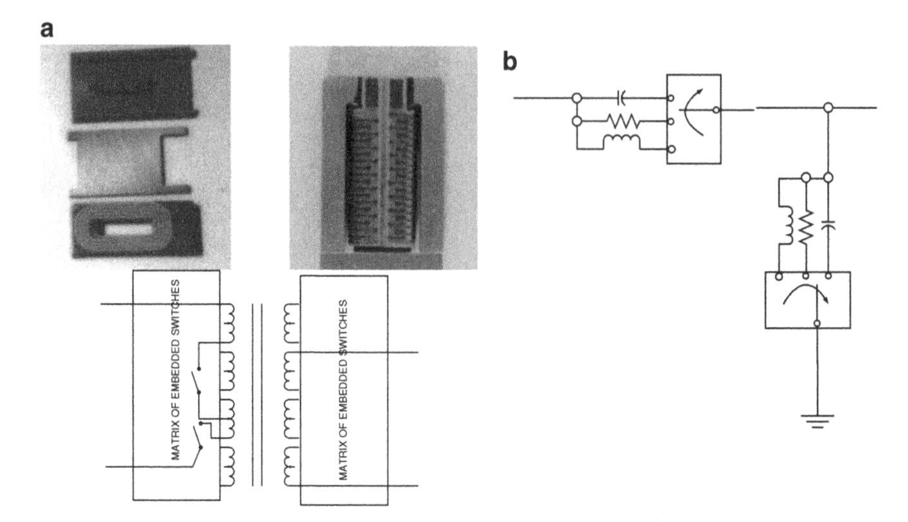

Fig. 13.4 Switched based reconfigurable analog elements. (**a**) Reconfigurable transformer. (**b**) Reconfigurable RLC filter

building blocks. This concept is shown in the examples of Fig. 13.4. The first example (Fig. 13.4a) illustrates a perspective configurable transformer, formed by using switches to effectively reconfigure the wiring pattern of one or both coils forming the transformer. In the second example (Fig. 13.4b), a more complex RLC filter circuit is shown in which individual components can be added or removed through switch closures, altering the effective transfer function of the filter.

A methodology for making reconfigurable capacitance is shown in Fig. 13.5. In this case, a micro-electromechanical systems (MEMS) device is created featuring louvers of a high permittivity material that can be altered in position using an actuator (not shown). As the louver is moved laterally, the effective permittivity is gradually increased, tuning the capacitor. Many other mechanisms could be devised for creating reconfigurable versions of basic building block elements.

More ambitious reconfigurable analog systems are commercially available in the form of programmable systems on-chip. As exemplified by companies such as Cypress Semiconductor (who offers products referred to as "programmable systems on-chip"), a contemporary product contains a mixture of configurable analog as well as digital intellectual property (IP) blocks, to include other diffused IP, such as microcontrollers.

13.1.2.5 Programmable Wiring

The idea of a programmable wiring system remains a future possibility in reconfigurable systems development. Some research had been undertaken at the Air Force Research Laboratory to develop a programmable wiring system capable of handling

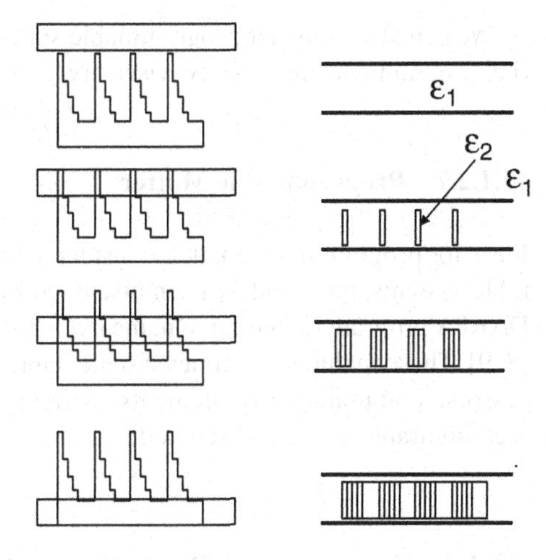

Fig. 13.5 Programmable permittivity based on prospective microelectromechanical systems (MEMS) device

a variety of power, weak signal analog, and microwave waveforms [5]. The project was referred to as the adaptive wiring manifold (AWM) or adaptive wiring panel (AWP). In this research, a mixture of integrated circuits and discrete components, including a large number of relays, were used to implement a configurable wiring system. The basic implementation strategy followed the concepts used in FPGA routing. The primary difference is that, whereas FPGA routing is limited to predominantly digital signals, the AWM was designed to handle power and analog signals.

Several generations of the AWM were explored. The earliest implementations employed latching, metal-to-metal MEMS switches. Early versions of the AWM were implemented on large circuit boards, with several hundred latching MEMS switches [6]. The early prototypes worked well, but despite the flexibility inherent in reconfigurable switching patterns, the boards were not scalable. To overcome this limitation, a cellular form of the AWM was created [7], in which a number of smart tiles formed the scalable fabric. Each cell was designed to connect to other tiles and contained 70 solid-state relays. A 48 cell demonstration system was constructed and successfully operated [1].

The notional idea of reconfigurable wiring should in fact be expanded to include a number of reconfigurable pathway concepts, including reconfigurable optical routing, and fluidic routing, thermal routing, and possibly other ideas involving different phenomenologies for which reconfiguration approaches might be applicable.

13.1.2.6 Programmable Mechanisms

Programmable mechanisms constitute a very broad class of potential reconfigurable systems, to include linear and rotational mechanical transport, modular robots, specialty motors and actuators, and elements of other configurable phenomenologies not previously discussed, to include reconfigurable sensors, and antenna.

We can also consider programmable surfaces and programmable bulk properties (i.e. programable emissitivity, viscosity, conductivity, etc.).

13.1.2.7 Programmable Matter

Ideas for programmable matter, considered to be a type of limit argument reconfigurable systems, have widely been discussed by computer scientists [10]. At least one DARPA program by the same name explored a number of eclectic research projects [8, 9]. These included centimeter-scale, fabricated cells comprised of actuators, processors, communication elements, and sensors, completely analogous to other reconfigurable systems discussed here.

13.1.3 Engineering Reconfigurable Systems

If we consider reconfigurable systems to be a field, then it is at best an embryonic field. Outside of the heavily developed discipline of the FPGA, there are few guiding principles governing the construction of future reconfigurable systems. We consider that there are several broad challenges in engineering systems to be reconfigurable, and these are briefly discussed.

13.1.3.1 Engineering Reconfigurability into System Components

A complex block of functionality might benefit from having a number of software adjustable “knobs”. We might call the art of designing components to have reconfigurable features “design for reconfigurability”. It involves careful consideration of what flexibility components should have, and how to engineer degrees of freedom in a way that is easily managed through a configuration mechanism.

13.1.3.2 Generalized Representation of Reconfigurable System

One general description of a reconfigurable system would be hierarchical. In this case a reconfigurable system would be considered a reconfigurable ensemble of components. The components themselves could be reconfigurable, and themselves could be a hierarchical arrangement of reconfigurable elements and so on. Fig. 13.6 provides a simplified depiction of a generalized reconfigurable system.

In this figure, a reconfigurable system is considered to consist of several reconfigurable modules, a programmable routing fabric, and configuration management engine. The details of these blocks are not shown, but consist of blocks engineered with configurable features whose settings are controlled by segments of the bitstream. The bitstreams of all configurable blocks, as well as those pertaining to

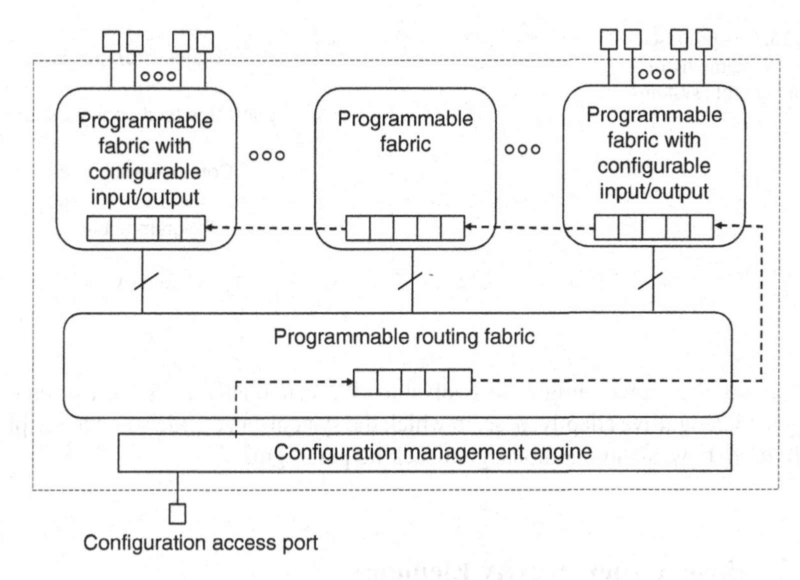

Fig. 13.6 Generalized representation of reconfigurable system

the programmable routing fabric, or managed by configuration management engine, are accessible through a configuration access port. The configuration access port is shown separate from other input/output pins. Some, but not all, configurable modules feature connections outside the system.

13.1.3.3 Adaptive Hierarchy

Reconfigurable systems expose knobs that can be manipulated to personalize them for specific functions. In principle, we can consider the philosophies for managing this collection of knobs, called the adaptive hierarchy. The adaptive hierarchy is depicted in Fig. 13.7 is a vertical progression. The bottom tier refers to a fixed system, which is the null or trivial case (i.e., there are no "knobs") Fixed-program (this is a configurable-one-time case, with a pre-compiled schedule of "knob" settings). Most reconfigurable systems occupied the next tier, simply referred to as "programmable". In this case, the schedules of "knob" settings are defined a priori. This is analogous to the case of precompiled software programs, which are both pre-written and processed or produced to run on a particular system, and FPGAs typically are synthesized in advance. A progress above a singularly programmed system is one that permits a number of pre-built configurations, in which a reconfigurable system is shifted from one configuration to another, which is referred to as *context switchable*. More ambitious possibilities can be considered. One possibility is self-configuration, in which a reconfigurable system can generate its own configuration given an input script. The situation is analogous to a system being able to fly its own toolchain. There is no reason for system to take on this level sophistication unless

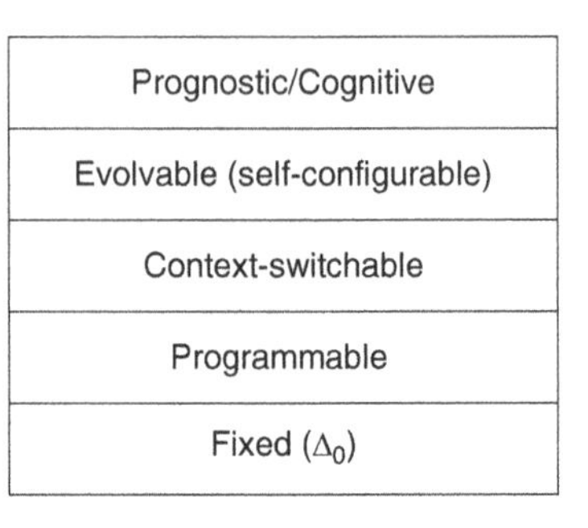

Fig. 13.7 A proposed adaptive hierarchy of reconfigurable systems

the scripts themselves might be evolvable. We can finally conjecture a type of prognostic/cognitive adaptiveness, in which the system can decide when to script, in addition to how, an autonomous, goal-seeking paradigm.

13.2 Basic Cybersecurity Elements

Cybersecurity broadly refers to protecting functions in systems (these being called "cyberphysical systems") that depend on the movement or configuration of information from deliberate (man-made) actions that could compromise them. The act of executing a set of actions (usually the modification of configurations inside embedded systems or the networks involving them) to produce these compromises is called a "cyber-attack". It can be as simple as setting a single bit in the register of a microprocessor from a one to a zero to produce an engine misfire, or it can be as sophisticated as the infamous StuxNet malware [11], which involved the use of multiple software modules (designed to execute on more than one type of operating system) coordinating the movement of malicious code (for subsequent execution) onto specific microcontrollers connected to a centrifuge for nuclear isotope separation.

Cyber-attacks have traditionally involved either: (1) network engineering, (2) embedded processor programming, or (3) combinations of the two [12] (in the next section we consider the extension of cyber attacks to the components and fabrics of reconfigurable systems).

Networks. As the backbones for data, networks provide the roadways for transmitting user information, command and control information, and potentially malicious content. The networks can be internal and external, multi-tapped or point-to-point, intentional and non-intentional.

Processors. Processors are complex state machines that run sequences of instructions. While most platforms have a distinct master or central computer, there are usually secondary and tertiary processors in most systems. For example, a personal computer has processors in the video graphics engine, disk controllers, and even in the keyboard and mouse. The basic signature of a cyber-attack involves the compromise of one or more of the instruction sequences in any of these processors through system code injection or pre-placed code insertion (during manufacture or original flight article preparation).

13.2.1 Examples of Cyber-Attacks on Embedded Systems

13.2.1.1 Stuxnet: Poster Child of Modern Cyber-Attack

Stuxnet in many ways represents the emergence of cyber-attack as a weapon. In the 2010 Stuxnet cyber-attack, a combination of techniques involving networks and embedded processors were used to deliver disruptive attacks on the centrifuges used in Iranian nuclear material preparation. The target in this case was an embedded controller (processor) contained within a piece of Siemens equipment that regulated the speed of the centrifuge. In order for the attack to work, a set of conditions had to exist conducive for exploitation: (1) a damageable action that could be effected through a program sequence on an embedded processor; (2) an ability to modify ("flash") the embedded processor; (3) a network connection from a host computer that could sequence communications packets to the embedded processor to implement the modification; (4) a malicious code pattern running on host computer that could invisibly perform the sequencing and spread itself on other hosts in its resident network; and (5) an "intrusion vector" for placing the malicious code on at least one machine in the network. Breaking any of these conditions would have rendered the attack ineffective.

13.2.2 Cyber Threat Effects Spectrum

Cyber threats are threats because of the effects that they can produce in their targets (sometimes referred to as victim) systems. Here, we briefly identify the obvious possibilities for malware.

13.2.3 System Defacing

Some cyber attacks can be undertaken for purely symbolic or political purposes. In these cases, the attacks are meant to convey a message, but in no other way produce a measurable effect or damage to the target system. They are nonetheless ominous, mostly due to the fact that they represent the ability of the author to infiltrate a system, presumably with the veiled threat of delivering a more serious impact.

13.2.3.1 System Degradation

The notion of system degradation is itself a spectrum of effects, ranging from the manifestation of small, almost immeasurable deviations and performance (that, for example, may eventually cause a system to drift from its intended performance baseline) to more dramatic denial of service and/or the unceremonious destruction of the system (i.e., "bricking").

13.2.3.2 Data Exfiltration

Sometimes the objective of malware is to extract information from a target system, either briefly or persistently, which in the latter case requires the engineering of malware to be minimally intrusive. The design for minimal intrusion includes not only making malware difficult to detect through inspection, but also masking its patterns of signaling and communication, that is the vectors through which it relays information to its author or user. The most convenient vectors for exfiltration are also the easiest to detect, namely those involving "in band" communications (i.e., relying on the same network channels used routinely by the system itself in its normal interactions).

13.2.3.3 System Repurposing

Malware can be used to hijack or repurpose the system, to exploit a system's normally intended purpose in ways other than intended by its creator. For example, perhaps a networked soda vending machine could be hacked to deliver free sodas with the right text message. In this case, the primary purpose of the malware is not to destroy or monitor a system, but rather to steal its products.

13.2.3.4 Accidental

There are possibilities, of a more academic nature, where malware is created for the purposes of study, and in some cases this malware can escape "into the wild". It may produce some of the above-mentioned effects, not through malicious intent, but as a byproduct. For example, self reproducing automata involving relocatable assembly code can easily be created to rapidly fill a memory space in a processor, quickly obliterating and overtaking any original code used by the processor for other functions. Sometimes these pedagogical mechanisms are themselves used as tools in a sort of malware toolbox, although their isolated study did not intend such purposes.

13.2.3.5 Symbiotic Malware?

Another possibility in malware is the design of code that actually produces no harm but coexists more or less harmoniously with other software. This idea is possibly a marginal corner case in the taxonomy of malware, and it does raise some definitional considerations, such as "What is malware?" We include the point mostly for completeness and in analogy of biological viruses. Although the origin of the word "virus" is synonymous with "poison", in nature a perfect virus is one that lives in harmony with its host. Sometimes the most destructive viruses are merely ones that have not evolved sufficiently. These viruses, over time (perhaps thousands to millions of years), would asymptotically become harmless to (if not symbiotic with) their hosts.

This is simply because by killing its host continued existence of a virus is compromised, and even absent of consciousness or intent, organisms tend to work in a way to propagate, and to prolong existence. Therefore, the "best" viruses are those that live to spread as pervasively as possible, meaning that the more hosts the better. It is perhaps nonsensical to try to apply such analogies to synthetic malware. Or conversely, perhaps, some if not all applications can be thought of as "harmless malware".

13.2.4 Malware Activation

Unless immediately triggered upon introduction, malware is likely dormant and must be activated through some means, and there are many creative possibilities. Possibilities may be grouped around two broad classes, namely external triggering and internal triggering.

13.2.4.1 Externally Triggered

External triggers involve conveying a signal from outside a system that results in the activation of cyber attack. There are many obvious vectors for external triggers, in the form of patterns or messages sent over networks. Less obvious are those triggers involving hidden receivers or involving sensors that supply some information or condition that results in the activation of malware. These can be subtle (such as a temperature drop involving a network thermometer) or sophisticated (such as using a particular image type or image feature).

13.2.4.2 Internally Triggered

Internal triggers involve the notion of a pattern in a state machine or some other threshold that causes a branch of code or other shift in operation resulting in the execution of malware. They can be widely varied, to include countdown timers, random numbers, or simply exploiting the knowledge of rare (but nonzero) probabilities of occurrence of code branches or other system events. While possibly less deterministic than external trigger, internal triggers can be used to create aging effects (malware that is likely to trigger only after certain span of time), to mimic decreased reliability (in larger populations of systems), or simply to exploit knowledge of particular system use cases to produce conditions likely to promote a trigger.

13.2.4.3 Signaling and Coordination in Distributed Malware

Signaling in complex malware, involving coordinated operation across many nodes or pieces of a system or network, can be thought of as an extension of triggering concepts.

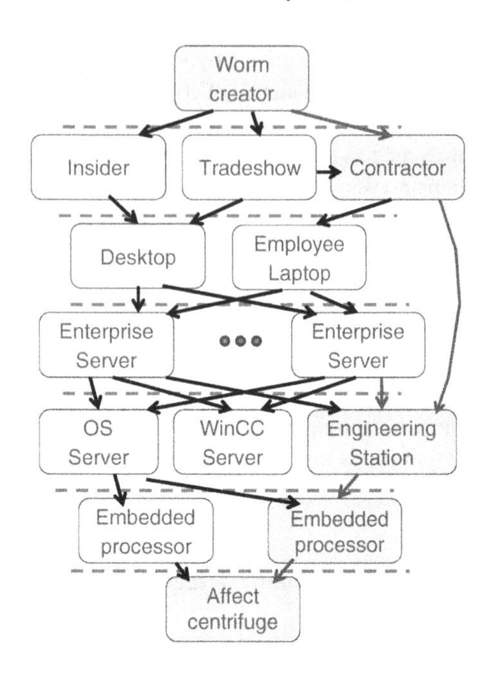

Fig. 13.8 Simplified attack graph (Stuxnet) after [15]

13.2.5 Understanding How Cyberthreats Can Affect Systems

How can cyber threats exist in our domains of concern?

In a sophisticated cyber-attack like Stuxnet [15], we see an effect produced as the goal of the attack (make the centrifuge spin dangerously fast over time), a series of vulnerabilities (easily reprogrammable controller, network path, and vulnerable host), and a set of conditions that are needed to exploit the vulnerability to achieve the intended effect. The combination of effects and vulnerabilities, if known exhaustively, can be analyzed to reveal potential exploits, which is sometimes called an attack graph [13, 14]. The attack graph can be used by an aggressor to find the easiest exploits or by a defender to verify there are no weak links (see Fig. 13.8).

13.2.6 Example of a Hackable Embedded System

It is instructive to consider a simple, albeit contrived, example cyber physical system (Fig. 13.9) to better understand the attack graph concept. This system represents generic platform consisting of several processors, sensor, an actuator, and a communications device.

The elements of the system are briefly defined as follows:

1. Actuator (simple thruster). This thruster is used to provide motion to the platform.
2. Interconnection between thruster an embedded processor.

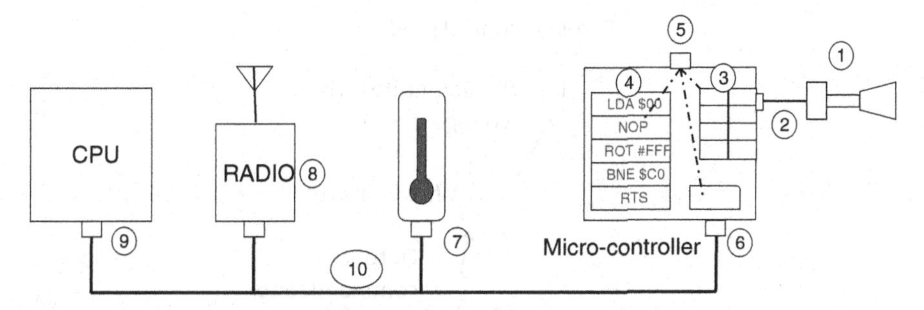

Fig. 13.9 Example cyberphysical system

3. Register file of embedded processor. Individual bits of selected registers are directly connected to general-purpose input/output (GPIO) lines, one of which directly drives (through a buffer amplifier) the #2 interconnection.
4. Program memory of embedded processor. This program member corresponds to the code space used by a dedicated microcontroller that operates the thruster under commands that come across the network.
5. Test and maintenance access port (1149.1 JTAG interface, for example). The JTAG interface allows low-level manipulation of registers, flip-flops, and memory stores throughout the microcontroller.
6. Connection to multidrop network. A multidrop network provides a robust connection to all nodes in the network. Some concepts here applied to more sophisticated switched packet networks.
7. Sensor (thermometer). A network device that reads platform temperature.
8. Communications device (radio). Provides communications between the platform and the outside world.
9. Central processing unit for platform. Provide centralized control for the platform overall
10. Multidrop bus

For this simple example, we define a thruster firing event as a possible target for a cyber attack (which is element #1). A simplified version of the corresponding attack graph for a thruster firing event shown in Fig. 13.10.

As suggested by the attack graph, there are many possible vectors for activating the thruster. It is ultimately only necessary that a voltage be set on a wire (2) connected to the thruster (3). By setting a particular pattern in a general-purpose input/output register (3), it is possible to form a voltage to achieve this. This can be achieved in turn through code manipulation (4), but also through direct interaction through an external JTAG port. Since the microcontroller in the thruster control module connects to a larger system through a multidrop bus (10), it is only necessary that a properly formulated network command be impressed upon the bus. The means for doing this can be as simple as implanting malware in the central processing unit (9) or as arcane as using a temperature-based trigger that activates malware in a heat sensing device on the network (7) that in turn produces a properly

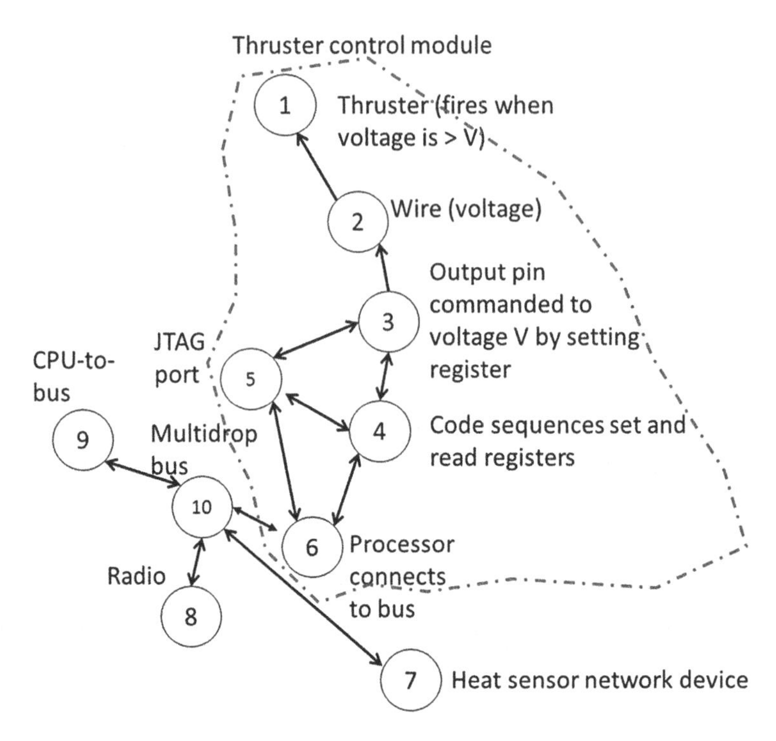

Fig. 13.10 Simplified attack graph

formulated network pattern. Of course, since the radio, which is also a network device, is connected to the same network, it also can become a convenient conduit for channeling commands directly on the network.

The utility of attack graphs is in eliminating the many possibilities for mobilizing a cyber attack on a system. It also provides some insights into the challenges of complete mitigation of such attacks. The JTAG port permits unfettered access to the internal logic circuitry and memory structures of the microcontroller. It is useful if not vital in troubleshooting complex systems. As shown in [16], such access ports can be used to compromise even secure systems. One means of mitigating this attack vector involves simply not instrumenting the access port in the final design. Ideally, this action nullifies node 5 in the attack graph. Through extensions of this basic method, it is possible to eliminate entire pathways for attack, thereby reducing the "attack surface".

It is also possible to modify the system design in a way that alters the attack graph, also ostensibly reduce the attack surface. Since all attacks must eventually result in the formation of a voltage that triggers a thruster firing, it is conceivable that one, for example, can place an interlock on the wire that is triggered through a separate means. If the interlock acts only as a switch, it cannot cause the thruster to fire, but can prevent a large number of other actions to fire it. Of course, if malware or some form of cyberattack affects the interlock, it can introduce the type of service denial, which is an entirely different type of problem.

13.2.7 Rudimentary Sketch of a Cyber Resiliency Approach

In summary, the attack graph can be used to formulate an enhanced awareness of cyber vulnerabilities and can also provide insight into their mitigation. A suggested, possible strategy includes the following elements:

- Develop a realistic set of attack graphs for the system of interest;
- Enumerate cyber attack scenarios based on the implied pathways for attack vectors;
- Formulate a rational cyber strategy that dramatically reduces the attack surface though a combination of:
- Eliminating unnecessary pathways for attack, by reducing unprotected configuration access ports or interfaces that allow key protections to be bypassed.
- Encumbering existing pathways through a variety of means, to include the introduction of blocks/interlocks, element encryption, intro element authentication, providing "out of band" approaches. It is important not to introduce changes to the attack graph that actually increase the attack surface

It is likely that no system can be perfectly protected against all cyber attacks, and it is unclear that it is possible to simply rank order a large number of vulnerabilities and attack them in order of decreasing probability. While on the surface this seems to be a reasonable approach, the attackers can also employ reasonable approaches and may choose attack vectors that are far more sophisticated and subtle. While to ignore glaring security flaws would border on irresponsible, one should not expect that possible yet improbable vectors might not be attempted.

As a type of reduction ad absurdum regarding the possible futility of finding all cyberattack vectors, we provide a simple example referred to as the "magic capacitor". A capacitor is a very simple and ubiquitous electronic component having two electrical terminals. They exist by the hundreds if not thousands on electronic circuit boards and boxes. We define a special type of magic capacitor that contains a complete electronic system, to include a tiny computer and a relay switch. The computer runs a software program generating random numbers repetitively. Upon encountering a specific combination, the computer closes a switch shorting the termini together. Since capacitors are commonly used for decoupling purposes, they are often placed between a voltage source and the ground for a circuit. Producing a dead short would obviously have a disastrous effect. Yet the existence of the magic capacitor might appear mysterious, since it is not accessible through a network, and is embedded in what appears to be an innocent component. A great many such examples, such as the magic capacitor, can be conceived. While an unlikely (improbable) attack vector, it is a possible attack vector. It would not likely be placed in an attack graph, any more than the totality of all imaginable cyber attacks could be placed in an attack graph. The example is not meant as an indictment against methods such as attack graphs, but merely to indicate that they (and likely most other conceivable methods) have limits.

13.3 Cybersecurity Challeges for Reconfigurable Systems

Reconfigurable systems extend the reach of cyber physical systems far beyond that of traditional embedded systems, and as such they become viable future opportunities for making software interact directly with the world around us. More specifically, the systems enable software defined hardware. This intimacy and deep connection between hardware and software is achieved through the notion of an accessible and reprogrammable configuration space, where bitstreams (or Boolean strings) define patterns of bits whose ones and zeros can control many software knobs and switches. As previously described, this configuration space allows the definition of the functions and eventually, perhaps, the structures of systems themselves.

Software knobs can be turned both for good and for evil. The configuration spaces of the future reconfigurable systems, in effect, become a playground in which opponents enjoy a potentially vastly increased attack surface. We can show, with simple examples some of the pathologies possible in future reconfigurable systems. It is, in effect, the future domain of cyber attacks for systems we, for the most part, cannot presently imagine, but are offered glimpses of through advanced research.

FPGAs. Field programmable gate arrays have advanced from simple devices that could replace a few scattered logic gates (such as the 22 V10 programmable array logic or "PAL" device) to sophisticated integrated circuit platforms capable of implementing millions of random logic gate equivalent designs, combined with implanted specialty finite state machines and in some cases multiple processing cores that are tightly coupled with the FPGA fabric itself. As price points have dropped below $20 per million gate equivalents, modern FPGAs have permeated into lower-cost electronic appliances, and eventually can be expected to be part of practically every consumer-electronics system.

In FPGAs, logic functions are implemented using lookup tables. A simple k-input lookup table (LUT), depicted in Fig. 13.11 (for $k=3$) is essentially a 2^k-bit configurable memory, capable of implementing any of the $2^\wedge(2^k)$ logic functions of k variables. The LUT is simply an addressing mechanism to access an individual bit of logic truth table. For the simple three input LUT (3LUT) in Fig. 13.11, a bitstream could map the functional specification for a 3-input AND gate, as shown in Fig. 13.12. Changing any single bit in the LUT changes the logic function.

The volatility of logic function due to changes in configuration memory space is illustrated in the simple example in Fig. 13.13. The generic 3LUT is capable of directly implementing the truth table of any 3-input logic function, as suggested in Fig. 13.13a. In this case, the y_i column is formed through the serial bitstream shown in Fig. 13.12. For example, a 3-input AND gate (3AND) is implemented by setting all memory bits to zero, except the one corresponding to $a=1$, $b=1$, and $c=1$. Any corruption in this bitstream changes the logic function. For example by setting the $i=4$ bit, the 3AND is converted into a 2AND, and the c input is effectively ignored.

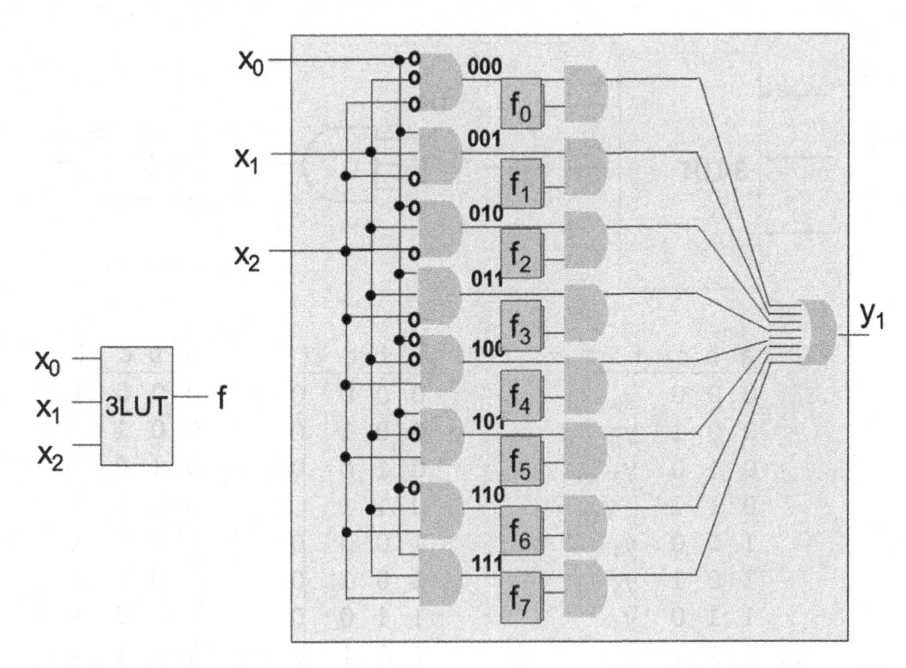

Fig. 13.11 *k*-input lookup table (*k*LUT) (*k=3*) and realization with logic gates and 2^k-bit memory

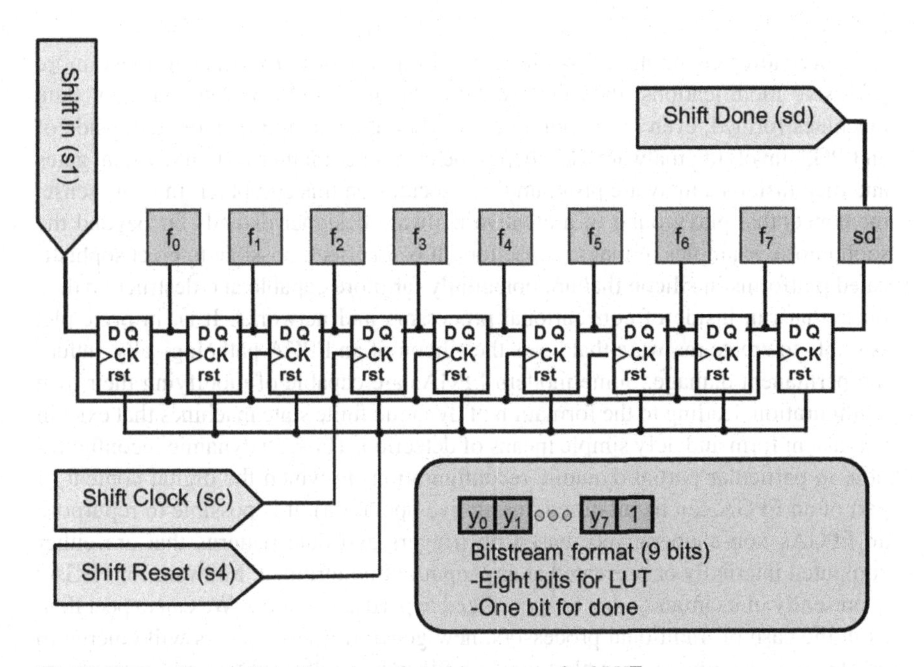

Fig. 13.12 Implementation of serial-configurable memory for *kLUT*

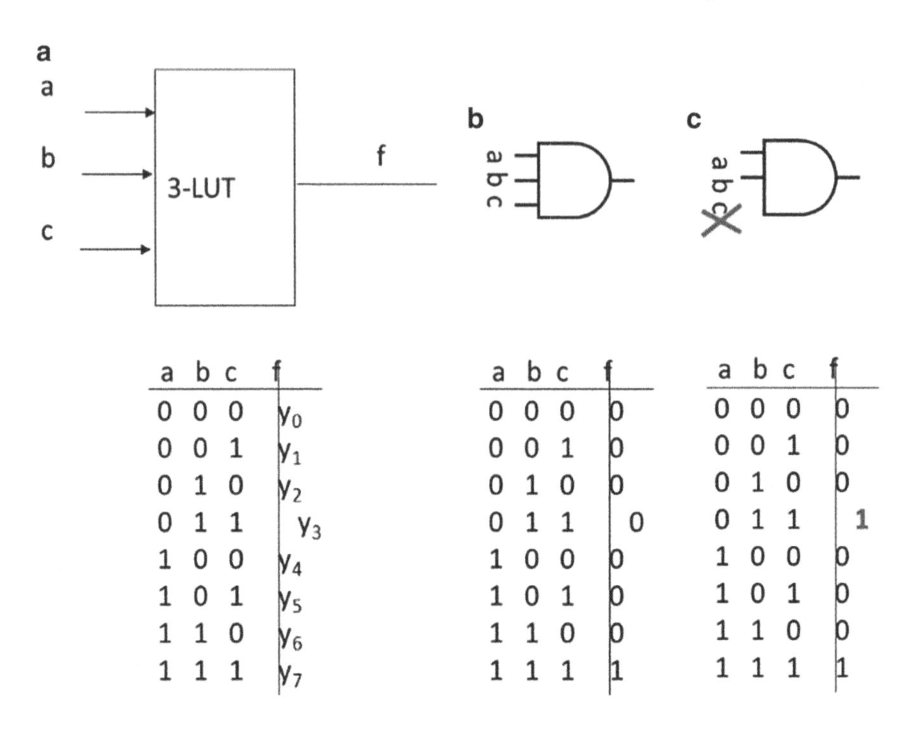

Fig. 13.13 Implementation and transmutation of logic functions in 3LUT. (**a**) Generic 3LUT. (**b**) Implementation of 3-input AND (3AND). (**c**) Mutation through bit change

If instead of changing a few bits in LUTs an opportunity exists to make more pervasive modifications, then clearly entire circuits could be rewired, new state machines formed, even to the point that a cyber attack could be mounted inside of an FPGA involving malware that shapes the custom computer from individual gates and then installs a malware program that executes on this computer. In some sense, the conceptual playground of a creative malware designer extends far beyond the sophomoric examples of magic capacitors. It is, in effect, possible to erect sophisticated platforms in silicon that are potentially far more capable and destructive than those that run inside of conventional processors and networks. It is, in principle, possible to create malware that heats the silicon of an FPGA pathologically, inducing permanent damage. Some modern FPGAs are capable of modifying their own configuration, leading to the formation of dynamic finite state machines that exist in a transient form and defy simple means of detection. Through dynamic reconfiguration, in particular partial dynamic reconfiguration (in which the digital content of part of an FPGA can be modified during live operation), it is possible to repurpose an FPGAs non-cooperatively based on triggers and data patterns that are either computed internally or presented at any input/output terminal. Malware for FPGAs is presently in its infancy, due to specialized expertise and tools. We can expect that, as in the case of traditional processors, new generations of hackers will emerge to create tools to empower exploitation of FPGAs in ways potentially even more profound than StuxNet.

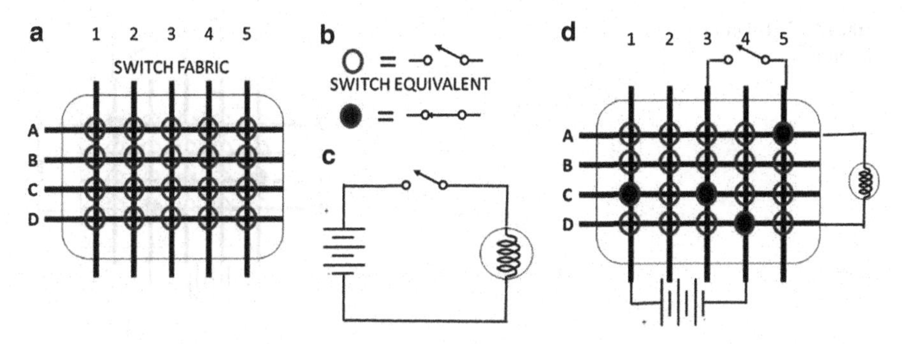

Fig. 13.14 A simple reconfigurable wiring harness. (**a**) Fabric definition. (**b**) Switch model. (**c**) Example circuit (flashlight). (**d**) Realization of flashlight with programmable harness

Other classes of reconfigurable systems, being mostly in the domain of research, are at a point in their evolution where it might seem that the consideration of cyber security is premature. We believe this is precisely the wrong view, as it is at the very point of conception for these new classes of systems that we have the most opportunities to engineer them responsibly and securely.

Consider for example, a simple future reconfigurable wiring harness (Fig. 13.14). The routing fabric that forms such reconfigurable wiring harnesses is analogous to those in FPGAs except that unlike FPGAs the resources in this wiring fabric are capable of supporting connections to analog, power, and microwave components. The particular scheme of this programmable wiring system is similar to a portion of an extended, cellular adaptive wiring system described in [7]. It consists of a fabric having a number of wire segments that can be connected together through embedded switches (Fig. 13.14a). The state of wiring is defined through a configuration management engine, that defines the state of a number of configuration switches (Fig. 13.14b). Completing the wiring pattern of a schematic only requires the closure of the correct switches. To implement a flashlight schematic (Fig. 13.14c), one connects each electrical component to particular terminus, generates the switch pattern (using a special-purpose compiler), forms a bitstream pattern, and conveys the bitstream pattern into the configuration management engine. The simple string defining the switch closure pattern in Fig. 13.14d is shown in Fig. 13.15.

Even this simple example can be a playground for cyber attack. The most obvious attack involves altering the bitstream which in effect rewires the physical system. We show several possibilities in Fig. 13.16. As we have seen in the study of radiation effects on FPGAs, many random bit flips can be harmless. Figure 13.16a depicts this case in which a randomly chosen bit is altered in the bitstream. However, a single bit alteration can also have a devastating impact it is shown in Fig. 13.16b in which the flashlight power source is short-circuited. Figure 13.16c furthers strains this contrived example system by adding a resistor externally and programming it to simply drain the battery faster. This corresponds to a case where a system is more subtly altered to degrade its performance. And of course the simplest attacks on the bitstream might be the cases where the entire bitstream is set to nulls, which simply disconnects the entire wiring system (Fig. 13.16d).

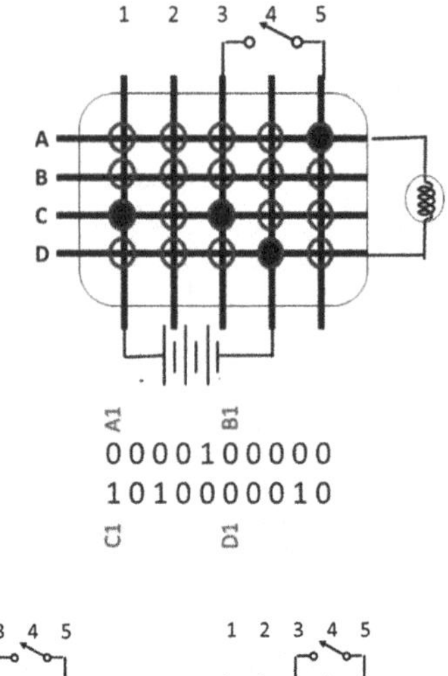

Fig. 13.15 Bitstream definition

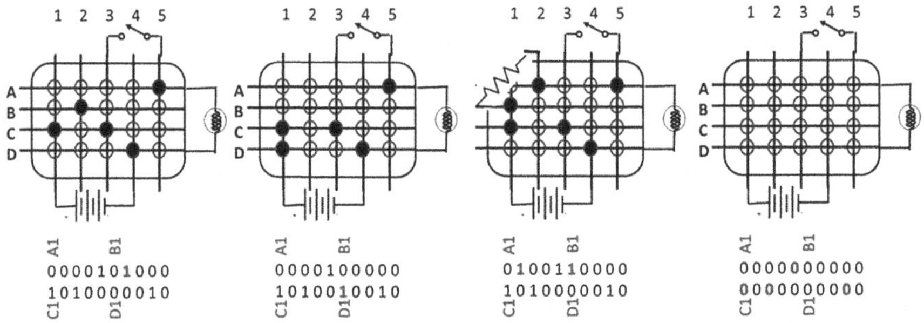

Fig. 13.16 Variety of bitstream modifications. (**a**) Random single bit change to bitstream. (**b**) Harmful single bit modification. (**c**) Mutation combined with external component. (**d**) "Zeroized" bitstream

13.4 Engineering Resilience in Reconfigurable Systems to Cyber Threats

The simple examples defined in the previous section do not represent a comprehensive resumé of possibilities for the extended reconfigurable systems of the future, but only begin to hint of the possibilities by which they may be compromised. In this section, we can at best motivate the need to consider resilience to cyberattack from the inception of the architectures that will serve as the basis of these systems.

13.4.1 Complicating Malicious Attempts to Reverse Engineer

While "security through obscurity" is fairly criticized as an inferior approach to protecting systems, particularly in isolation of other methods, it can, with care, provide an effective complement. For example, methods of code obfuscation have long been employed in software design, and the hardware equivalents of these can also be considered. Intellectual property (IP) blocks can effectively be encrypted through obfuscation, at least creating a substantial work effort barrier. Communications channels in networks can be designed similarly to obscure traffic patterns with the goal of providing iso-entropic information content. Such designs make it difficult for attackers to glean useful insights inside channel analysis. Ideally, these approaches can be in some way “decloaked” for developers, so that productive development and troubleshooting activities are not encumbered by the obscuration processes.

13.4.2 Designing for Resilience

In the ground floor of the design of architectures for sophisticated reconfigurable systems of the future, a number of techniques can be considered that while somewhat complex are far simpler to implement than after-the-fact techniques. These include channel encryption and inter-component authentication. Obviously, channel encryption (in complement to traffic obscuration) not only renders the patterns of communications unintelligible but also makes the contents indecipherable. Through component authentication, protocols are established were components within networks can in effect “log into” each other. Through these approaches, it is possible to establish multilevel security enclaves within a larger system. Components that do not have the necessary keys or do not understand the associated protocols are unable to join the network, a method that is made more powerful when communication channels are also encrypted.

These approaches can be further bolstered by employing split channel methodologies. For example, components with multiple network ports might diversify transactions by randomly splitting transactional information across each.

13.4.3 Awareness and Trust of Components and Supply Chains

Supply chains remain a large challenge to securing modern systems. They include not only the stewardship of a system’s bill of materials (the literal monitoring of all physical piece parts that make up a system), but also the contents of all digital programmations to include not just traditional software programs, but IP blocks, permanent and transient data structures, and configuration spaces to include those of reconfigurable devices. We must not only trust the products, but the processes that are used to create them.

An element of trust involves establishing the pedigree of components and intellectual property, and a number of ideas are being discussed to achieve this. While being able to establish proof of origin does not guarantee a system is malware-free (any more than we can guarantee that a system is defect-free), the ability to trace origination of the elements of systems reduces unknowns and helps build confidence.

Some of the more intriguing ideas include the embedding of taggants (which amount to traceable nanoscale "bread crumbs") in hardware, the exploitation of physically unclonable functions (PUFs), and, for software, establishing methods that embed custody chains (such as driver signing/certification). These approaches help assist in the detection of counterfeit components, provide means for tracing back some aspects of component origin, and, for software, may additionally provide checks on the integrity of digital construction.

13.4.4 Trusted Autonomous Embedded Monitor/Controllers (TEMC)

The trusted autonomous embedded Monitor/controller (TEMC) can be thought of as a micro-black recorder, although a building intrusion alarm may be a better metaphor. The TEMC is defined as a separate microcontroller-based circuit placed within an embedded system, which is a client for the TEMC function.

The TEMC concept is shown in Fig. 13.17, using the system from Fig. 13.9 as an example client. The TEMC is a self-contained system, containing a processor (A) and several special network ports. These network ports are routed throughout the client system to service monitor or control points working under a software application running on the TEMC processor. The network is a series of point-to-point links, connecting to special endpoints. The endpoints employ a uniform interface to the TEMC, but may employ customized circuitry according to purpose. Fig. 13.17 depicts several types of endpoints. The first of these are specially designed, processor control endpoints (B). These endpoints are designed to connect to access ports in particular processors within a client system. The capabilities are varied according to the specific processor they are connected to. In this example, the endpoint connected to the central processing unit (CPU) involves a proprietary interface for which an endpoint has been custom-designed. The endpoint connected to the second embedded controller (connected to the thruster) employs the JTAG interface previously discussed. Both of the (B) type endpoints have the ability to do deep inspection of internal states of processors, as well as conducting overrides to include restoring software to a known state (for example, using backup copies stored within the TEMC). Another type of endpoint (C) is capable of monitoring network traffic. A more sophisticated form, suggested here, is capable of disconnecting components from the network and, furthermore, overriding signals on the network. In this case, the CPU can be effectively severed from the network, allowing the TEMC to drive the network directly. Since the TEMC might be a less capable processor, such overrides might be considered an emergency measure allowing the system some limp-along/recovery capability in the case of a severe compromise of the CPU due to hardware failure, malware, or other software defects. Another type of endpoint shown is an application-specific design (D), in this case one capable of blocking a signal going to the thruster. The TEMC can implement a series of rules that keep the thruster disconnected from the rest of the system unless the rules are satisfied.

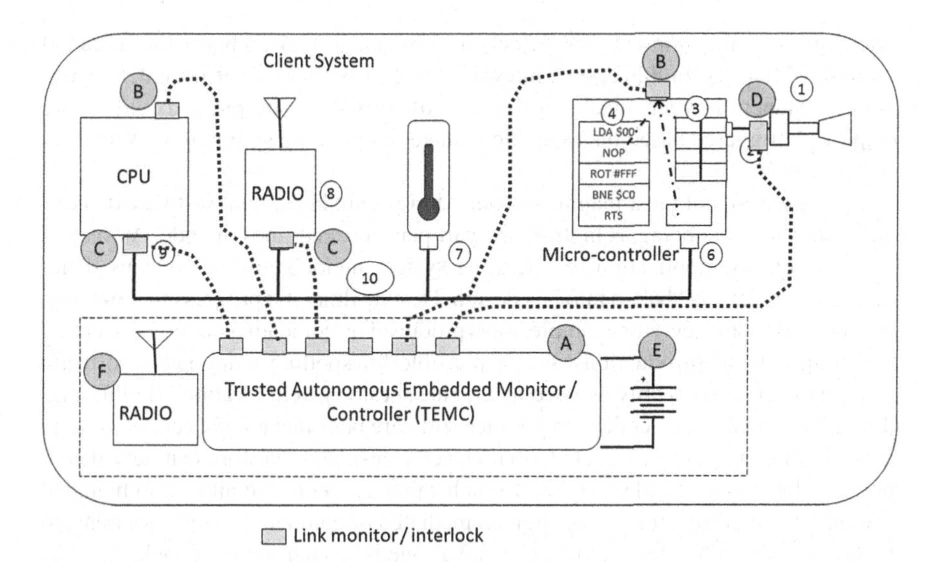

Fig. 13.17 Trusted autonomous embedded monitor/controller (TEMC) concept

This amounts to a hardware enforced type of white listing, one that cannot easily be tampered with from the client system.

TEMCs, such as shown in Fig. 13.17, can be constructed by observing a few key principles:

- Implementation is performed independently of the client system, under the notion of a "root of trust". The TEMC must not be subject to the same types of corruption that might afflict the clients it attaches to
- Ability to "air gap" TEMC from networks of client, making it autonomous
- Ability to self-power (trickle charge from the client if necessary)
- Ability to communicate over an independent network to an administrative agent that is trusted with oversight of the TEMC and works closely with administrative agent for the client.

13.5 Is There a Scientific Basis for Cyber-Immunity?

It is attractive at one level to think there might be some basis in science for deciding that a particular embedded system can be made immune to cyber attack. We briefly consider this problem, and we unfortunately conclude that there is little hope for a breakthrough that makes a system immune to attack through cyber means.

Malware is comparable to software defects in that both are undesired occurrences. The primary difference is that malware is man-made, whereas software defects are accidental. Both can cause departures in desirable behavior. In the one case, the departure from desired behavior is an engineered effect. In fact, malware often builds on the unintentional defects in software systems, namely the so-called "zero day" defects. In this case, a malware design harnesses a software defect as a hook to launch a cyber attack. In the case of StuxNet, multiple zero days were employed, triggered based on a specific vintage of operating system (e.g., Windows XP vs. Vista).

Even the best and most mature software designs are laden with software defects, the occurrence rate being as high as one part per thousand lines of code. As such, a multimillion line application or operating system undoubtedly has dozens if not thousands of defects "baked into" its design. Despite dramatic progress in tools such as static code analyzers (for example, tools produced by companies such as Coverity), which upon their introduction were responsible for spotting many latent software bugs, static analysis is only as good as the rule set it is given. "Nature" (in this case the nature of man-made code) can produce software bugs that are not caught in static analysis. By analogy, we conclude that cleverly designed malware can defy detection, can be "hidden in plain sight". In such cases, it seems that almost no bounded amount of code inspection, either human mediated or automated, can be guaranteed to discover all such defects. And it only takes one to launch a cyber attack.

In this respect, reconfigurable systems do not change this game, but at one level—pessimistically—they only increase the attack surface. But, just as adding lines of code in software also decreases reliability, it is important for the sake of progress and advancement that we will make more complex and capable systems, balancing these risks. At best, our objective in engineering cyber resilient reconfigurable systems is to minimize this attack surface by thoughtful engineering.

References

1. Victor Murray,,Daniel Llamocca, James Lyke, Keith Avery,Yuebing Jiang, Marios Pattichis."Cell-based Architecture for Adaptive Wiring Panels: A First Prototype", accepted, Journal of Aerospace Information Systems, Vol 10, No. 4, April 2013.
2. Sipser, Michael. Introduction to the Theory of Computation. Cengage Learning; third edition, 2012.
3. Hartenstein, Reiner W. and Herbert Brunbacher (editors), Proceedings of Field Programmable Logic and Applications: the Roadmap to Reconfigurable Computing., Lecture Notes in computer science, vol 1896, Spring-Verlach, New York, 2000.
4. Haykim, Simon. Neural Networks and Learning Machines, Third Edition. Prentice Hall, 2008.
5. Wilson, Warren, Jim Lyke, and Paul Contino,"MEMS-based Reconfigurable Manifold", presented at MAPLD 2001, Johns Hopkins University- Applied Physics Laboratory, September 11-13, 2001.
6. Warren Wilson, James Lyke and Glenn Forman, "MEMS-based Reconfigurable Manifold Update", Proceedings of the MAPLD Conference, September 7-9, 2005, Washington, DC.
7. Victor Murray, Daniel Llamocca, Yuebing Jiang, James Lyke, Marios Pattichis, Stephen Achramowicz, Keith Avery, "Cell-based Architecture for Adaptive Wiring Panels: A First Approach", Proceedings of Reinventing Space Conference (RS2011), 2-6 May 2011, Los Angeles, CA.
8. E. Hawkes, B. An, N. M. Benbernou, H. Tanaka, S. Kim, E. D. Demaine, D. Rus, and R. J. Wood, Programmable matter by folding, PNAS 107 (28) 12441-45 (2010).

9. K. Gilpin, K. Ara, R. Daniela, Robot Pebbles: One Centimeter Modules for Programmable Matter through Self-Disassembly, IEEE International Conference on Robotics and Automation (ICRA), 2485-92, (2010), Eds. M Rakotondrabe and I. A. Ivan.
10. B. J. MacLennan, Universally programmable intelligent matter summary, Proceedings of the 2002 2nd IEEE Conference on Nanotechnology, 405—08 (2002).
11. D. Kushner, The Real Story of Stuxnet, IEEE Spectrum 50 (3) 48-53 (2013).
12. C. Schuett and J. Butts, An evaluation of modification attacks on programmable logic controllers, Int. J. Crit. Infrastructure Protection, 7 (1), 61-8 (2014).
13. N. Idika and B. Bhargava, Extending Attack Graph-Based Securith Metris and Aggregating Their Application, IEEE Trans. Dependable and Secure Computing 9 (1), 75-85 (2012)
14. C. Phillips and L.P. Swiler, A Graph-Based System for Network-Vulnerability Analysis, NSPW'98: Proc. Workshop New Security Paradigms. pp. 71-79, 1998.
15. Byres, Eric; Ginter, Andrew; and Joel Langill, "How Stuxnet Spreads – A Study of Infection Paths in Best Practice Systems", white paper, Tofino Security (http://ScadaHacker.com), February 22, 2011.
16. Sergei Skorobogatov, Christopher Woods. "Breakthrough Silicon Scanning Discovers Backdoor in Military Chip", Springer-Verlag, Cryptographic Hardware and Embedded Systems – CHES 2012,Lecture Notes in Computer Science Volume 7428,Proceedings of the 14th International Workshop, Leuven, Belgium, September 9-12, 2012 pp 23-40.

Printed in the United States
By Bookmasters